S0-DJM-750

Globalizing Integrated Pest Management

A Participatory Research Process

edited by

George W. Norton

Department of Agricultural and Applied Economics,
Virginia Polytechnic Institute and State University,
Blacksburg, Virginia, U.S.A.

E. A. Heinrichs

Office of International Research, Education, and Development
Virginia Polytechnic Institute and State University,
Blacksburg, Virginia, U.S.A.

Gregory C. Luther

The World Vegetable Center
Taiwan

Michael E. Irwin

Entomology Department
University of Illinois
Urbana-Champaign, Illinois, U.S.A.

Globalizing Integrated Pest Management — A Participatory Research Process, edited by George W. Norton, E.A. Heinrichs, Gregory C. Luther, and Michael E. Irwin

Blackwell Publishing Professional
2121 State Avenue, Ames, Iowa 50014, USA

Orders:	1-800-862-6657
Office:	1-515-292-0140
Fax:	1-515-292-3348
Web site:	www.blackwellprofessional.com

Blackwell Publishing Ltd
9600 Garsington Road, Oxford OX4 2DQ, UK
Tel.: +44 (0)1865 776868

Blackwell Publishing Asia
550 Swanston Street, Carlton, Victoria 3053, Australia
Tel.: +61 (0)3 8359 1011

This book was produced from camera-ready copy supplied by the authors.

First edition, 2005

ISBN-10: 0-8138-0490-6
ISBN-13: 978-0-8138-0490-3

Library of Congress Cataloging-in-Publication Data available on request

The last digit is the print number: 9 8 7 6 5 4 3 2 1

Contents

III. Deploying Strategic IPM Packages

IV. Evaluating Strategic IPM Packages

V. Conclusions

Contributing Authors

Jeffrey Alwang, Dept. of Agricultural and Applied Economics, Virginia Tech, USA

Linda Asturias de Barrios, Estudio 1360, Guatemala

Aurora Baltazar, University of the Philippines–Los Baños, Philippines

Victor Barrera, Instituto Nacional de Ciencias Agropecuarias (INIAP), Ecuador

Darrell Bosch, Dept. of Agricultural and Applied Economics, Virginia Tech, USA

John Caldwell, Japan International Centre for Agricultural Science (JIRCAS), Japan

Philip Chung, Rural Agricultural Development Authority (RADA), Jamaica

Dionne Clark-Harris, Caribbean Agricultural Research and Development Institute (CARDI), Jamaica

Charles Crissman, Centro Internacional de la Papa (CIP), Kenya

Leah Cuyno, Northern Economics, Inc., USA

Lefter Daku, Dept. of Agricultural and Applied Economics, Virginia Tech, USA

Thomas Debass, OPIC, USA

S.K. De Datta, Office of International Research, Education, and Development, Virginia Tech, USA

Clive Edwards, Department of Entomology, Ohio State University, USA

Michael Ellis, Dept. of Plant Pathology, Ohio State University, USA

J. Mark Erbaugh, Inter. Programs in Agriculture, Ohio State University, USA

Bardhosh Ferraj, Fruit Tree Research Institute, Albania

Shelby Fleischer, Dept. of Entomology, Penn. State University, USA

Kevin Gallagher, FAO Global IPM Facility, Rome

Kadiatou Touré Gamby, Institut d'Economie Rurale (IER), Mali

Victor Gapud, Dept. of Entomology, University of Philippines–Los Baños (UPLB)

Sarah Hamilton, Dept. of Anthropology, University of Denver, USA

Colette Harris, Office of International Research, Education, and Development, Virginia Tech, USA

Myzejen Hasani, Agricultural University of Tirana, Albania

E.A. Heinrichs, Office of International Research, Education, and Development, Virginia Tech, USA

K.L. Heong, International Rice Research Institute (IRRI), The Philippines

Michael E. Irwin, Dept. of Entomology, University of Illinois, USA
Michael Jackson, U.S. Dept. of Agri. (USDA)/Agri. Res. Service (ARS), USA
James Julian, Hort. and Landscape Arch., Purdue University, USA
A.M.N. Rezaul Karim, Bangladesh Agricultural Research University (BARI), Bangladesh
Samual Kyamanywa, Dept. of Crop Science, Makerere University, Uganda
Janet Lawrence, Rural Agricultural Development Authority (RADA), Jamaica
Gregory C. Luther, The World Vegetable Center, Taiwan
James Mangan, Consultant, Australia
Sally Miller, Dept. of Plant Pathology, Ohio State University, USA
Keith Moore, Office of International Research, Education, and Development, Virginia Tech, USA
George W. Norton, Dept. of Agricultural and Applied Economics, Virginia Tech, USA
Joseph Ogrodowczyk, Graduate Student, University of Massachusetts, USA
Douglas Pfeiffer, Department of Entomology, Virginia Tech, USA
David Quishpe, Instituto Nacional de Ciencias Agropecuarias (INIAP), Ecuador
Edwin G. Rajotte, Dept. of Entomology, Penn. State University, USA
Agnes Rola, University of the Philippines–Los Baños, Philippines
Carolyn Sachs, Dept. of Agricultural Economics and Rural Sociology, Penn. State University, USA
Guillermo Sánchez, Denominada Inst. C.A. de Desarrollo Agro. (ICADA), Guatemala
Steven Sherwood, World Neighbors, Ecuador
Brunhilda Stamo, Plant Protection Institute, Albania
Carmen Suarez, Instituto Nacional de Ciencias Agropecuarias (INIAP), Ecuador
Glenn Sullivan, Hort. and Landscape Arch. (retired), Purdue University, USA
Irene Tanzo, Dept. of Agricultural Economics and Rural Sociology, Penn. State University, USA
Daniel B. Taylor, Dept. of Agricultural and Applied Economics, Virginia Tech, USA
Josef Tedeschini, Plant Protection Institute, Albania
Jessica Tjornhom, State Street Investments, USA
Sue Tolin, Plant Pathology, Physiology, & Weed Science Dept., Virginia Tech, USA
Rexhep Uka, Agricultural University of Tirana, Albania
Steven Weller, Horticulture and Landscape Arch., Purdue University, USA
Roger Williams, Dept. of Entomology, Ohio State University, USA
Takayoshi Yamagiwa, Graduate Student, Virginia Tech, USA

List of Figures

List of Tables

Acknowledgments

The editors and authors thank the 100+ scientists (host country, U.S., and international agricultural research centers), research administrators, NGO representatives, private sector representatives, and students who have collaborated with the IPM CRSP program in various countries over the past ten years. Many of them are cited individually at the end of the regional chapters. We also thank the hundreds of farmer collaborators in the various research sites. We appreciate the contributions of current and former IPM CRSP external evaluation panel members Don Plucknett, Sonny Ramaswamy, Doug Rouse, Mel Blase, Shelley Feldman, and Laurian Unnevehr. We thank former IPM CRSP program director Brhane Gebrekidan, as well as the many current and former Board members who contributed. The efforts of the entire IPM CRSP and OIRED staff — especially Debbie Glossbrenner, Peggy Lawson, Gene Ball, Larry Vaughan, and Miriam Rich — are greatly appreciated.

The editorial and production assistance of Mary Holliman, Bruce Wallace, and Leigh Corrigan of Pocahontas Press was invaluable in producing this book. The editorial assistance of Antonia Seymour and Dede Anderson at Blackwell Publishing is also gratefully acknowledged. The financial support of USAID/EGAT (Grant No. LAG-G-00-93-0053-00), and the technical and administrative guidance of the IPM CRSP CTO Bob Hedlund have been indispensable.

Finally, the editors give their greatest thanks to Dr S. K. De Datta, Principal Investigator of the IPM CRSP. His tremendous knowledge, energy, and fair and transparent style of managing the IPM CRSP has been the driving force behind the success of the project on which much of this book is based.

Foreword

S.K. De Datta,
Principal Investigator IPM CRSP, Director OIRED, and
Associate Provost International Affairs,
Virginia Polytechnic Institute and State University

It is generally agreed that the research, education, and development enterprise in the United States is the strongest in the world. However, all industrialized countries and some newly industrialized developing countries, such as China and India, are also engaged in cutting-edge science and technology for economic development in their countries. It is critical that U.S. universities engage in a global agenda of collaborative research, education, and outreach with other countries, not just for the benefit of those countries, but for U.S. national well-being as well, due to synergies and mutual gains from these efforts. The topic of this book, integrated pest management, is one of the areas in which international collaboration is crucial and, potentially, mutually beneficial. Pests have little respect for borders, and producers and consumers at home and abroad stand to gain from increased efforts to develop and implement IPM strategies around the world.

The U.S. Agency for International Development (USAID), through its Economic Growth and Trade (EGAT) Bureau, has championed global collaboration in agricultural research by providing resources to nine Collaborative Research Support Programs (CRSPs) with annual funding of approximately $22 million. USAID regional bureaus and missions around the world have complemented USAID EGAT funding with additional support. One of the CRSP programs, initiated in 1993, is the Integrated Pest Management (IPM) CRSP. The IPM CRSP was conceptualized by USAID to address health, environment, and economic issues globally through IPM interventions. The IPM CRSP, awarded competitively to and managed by Virginia Tech, has taken a consortium approach involving other U.S. universities, international agricultural research centers (IARCs), non-governmental organizations (NGOs), and private sector entities to plan, develop, and transfer successful IPM technologies. Many of the examples found in this book present and draw on lessons learned on the IPM CRSP.

The development and transfer of IPM technologies depend on a combination of technical and socioeconomic factors. In the IPM CRSP, we believe that a participatory approach should be followed in all aspects of program management, problem identification in targeting crops and pests, development of new technologies, and technology transfers to end users. The stakeholders in these participatory processes include, among others, farmers, farmer cooperatives, scientists, teachers, bankers, pesticide companies, marketing agents, policy-makers, and extension agents. The products of these processes on the IPM CRSP include the development and institutionalization of eight regional IPM programs in Africa, Asia, Latin America, Caribbean, and Eastern Europe. Relevant new cost-effective technologies; printed materials — such as scientific articles, books, fact sheets, symposia, and conference proceedings; and human capital developments — such as the training of farmers, extension agents, and researchers in short-term and long-term programs (the latter including the training of graduate students) are among the outputs. Throughout the process, progress was continually monitored and impacts were measured against benchmarks. Policies that affect adoption or non-adoption are also critical elements, and the IPM CRSP has evaluated numerous policy instruments.

Gender equity concerns are not merely academic, but are critical issues that impact crop production and the heath and safety of household members. IPM interventions, such as the minimal use of safe pesticides for maximum benefit, are often constrained by not sufficiently engaging women stakeholders. The IPM CRSP has developed methods for integrating gender issues into the IPM implementation process. The IPM CRSP has also used simple IPM interventions as well as cutting-edge scientific tools, such as biotechnology and GIS, in developing knowledge and technologies for end users.

The Office of International Research, Education, and Development (OIRED), which has served as the Management Entity, has attempted to manage the IPM CRSP as openly and fairly as possible, devoting significant Virginia Tech resources to ensure its success. Collaborative partners in the United States, which also have devoted significant resources to ensure project success, include: Penn State, Ohio State, Purdue, U.C.-Davis, University of Georgia, Montana State, the U.S.

Department of Agriculture, and historically black colleges and universities such as N.C. A&T, Fort Valley, Florida A&M, Maryland-Eastern Shore, and Lincoln. Numerous host-country institutions, international agricultural research centers (especially AVRDC, CIP, and IRRI), and NGOs in all regions have collaborated with U.S. scientists and have provided laboratory, field, and office space for the project. USAID missions in Ukraine, Albania, Uganda, Mali, Guatemala, and the Caribbean region have provided about $5 million for IPM activities that relate to Mission priorities.

This book should serve as an important resource for all IPM practitioners as well as for domestic and international development agencies such as USAID and development banks. The technical editors and the authors of the chapters in this book have written from first-hand knowledge and experience gained from serving in IPM programs, especially the IPM CRSP. We appreciate the contributions to Chapter 9 of Kevin Gallagher of the FAO Global IPM Facility; Steve Sherwood, who serves as Andean area representative for World Neighbors in Ecuador; and James Mangan, who worked with FFS programs in Indonesia and China. We, the IPM CRSP family, are grateful to USAID/EGAT for funding the IPM CRSP and to the IPM CRSP CTO, Dr. Robert C. Hedlund, for his leadership in the project. We believe this compendium will be a valuable addition to the literature on global IPM issues.

Figure 1-2. Ugandan farmers participating in sorghum IPM training.

Figure 2-4. Eggplant IPM experiment in farmer's field in the Philippines.

Figure 3-1. Fruit and shoot borer Leucinodes orbonalis *damage to eggplant.*

Figure 3-3. Purple nutsedge (Cyperus rotundus) *and* Trianthema portulacastrum *in onion in the Philippines.*

Figure 4-1. Striga in bloom in Mali.

Figure 5-2. Late blight (Phytophthora infestans) *on potato.*

Figure 5-5. Recording data from plantain experiment in Ecuador.

Figure 6-4. IPM scientist with pheromone trap in Jamaica.

Figure 8-2. Farmer field day in Bangladesh.

Figure 9-2. Neem tree in Mali

Figure 9-3. Farmer Field School in the Philippines.

Figure 11-2. Snowpea cold room storage in Guatemala.

Figure 11-3. Packing green beans for export in Mali.

Figure 13-1. Farmer spraying insecticides on onions in the Philippines.

Figure 14-1. Women farmers in India.

Figure 14-2. Women sorting out insect-damaged potatoes in Ecuador.

— I —
The Need to Globalize IPM through a Participatory Process

— 1 —
The Need for Cost-Effective Design and Diffusion of IPM

George W. Norton, S.K. De Datta, Michael E. Irwin,
Edwin G. Rajotte, and E.A. Heinrichs

As food demand has grown worldwide, agricultural production has intensified and there has been a concomitant expansion in the use of synthetic pesticides. Questions are increasingly raised about the sustainability of production systems heavily dependent on such chemicals. Concerns over potential health and environmental dangers, increased pest resistance to pesticides, and continued prevalence of pest-induced crop

Figure 1-1. Farmer and scientists discussing potato insect problems in Ecuador.

losses, especially in areas where pesticides have not reached, are stimulating the search for strategies that utilize genetic, cultural, biological, information-intensive, and other pest management alternatives.

Integrated pest management (IPM) is becoming increasingly important globally for managing agricultural pests, especially in more developed countries. IPM is a management system philosophy that emphasizes using increased information to make pest management decisions and integrating those decisions into ecologically and economically sound production systems. It utilizes all suitable techniques and methods in as compatible a manner as possible to maintain pest populations below those causing economic injury (FAO, 1967). For example, components of IPM systems include biological, environmental, and economic monitoring, predictive models such as economic thresholds, and a variety of genetic, biological, cultural, and, when necessary, chemical control measures. However, IPM efforts are still heavily concentrated in developed countries, despite intensifying chemical use in many developing countries as well.

The past 20 years have witnessed a growth in IPM programs in selected developing countries such as Indonesia, the Philippines, China, Uganda, and Guatemala, but adoption of IPM remains slow in most of the developing world. The International Agricultural Research Centers (IARCs), the Food and Agricultural Organization (FAO) of the UN, the U.S. Agency for International Development (USAID), CARE, and other organizations have collaborated with scientists and extension workers in developing countries to encourage the development and deployment of IPM systems. Adoption is limited, however, due to technical, institutional, social, cultural, economic, educational, informational, and policy constraints. Future expansion of IPM in developing countries will depend on success in reducing each of these constraints. Ultimately, the adoption of IPM strategies rests in the hands of farmers, as pest management practices must meet their needs as well as those of society. Stronger science must be melded with more cost-effective farmer-participatory approaches. The combination of good science and cost-effective participatory research and training has proven elusive in many IPM programs. In addition, the appropriate role of the public versus the private sector remains ill-defined, particularly so for the development and implementation of newer IPM tools such as biotechnology.

Multiple Approaches and Tradeoffs

A variety of approaches are available for developing and extending IPM solutions to pest problems. They include traditional research and extension methods that rely heavily on laboratory/green house and on-station research combined with extension education (public and private) through individual contacts, meetings, publications, and electronic media; on-farm structured experiments and farming systems research and training; participatory "farmer-field-schools" that emphasize research and education programs that rely heavily on farmer-generated research and training; and variants of these approaches.

Emphasis in IPM programs has gradually shifted to more participatory approaches (Van de Fliert, 1993; Dlott et al., 1994; Röling and Van de Fliert, 1994; Norton et al., 1999). IPM is particularly suited to participatory research and extension (R&E) because it is multidisciplinary and management- and information-intensive. Finding solutions to pest problems is high on the agendas of farmers, and farmers often have significant pest management knowledge and occasionally like to experiment (Bentley et al., 1994).

Participation, however, can have many goals (empowerment, technology generation and diffusion) and interpretations, and a broad array of participatory methods have been applied in R&E programs (Thrupp and Haynes, 1994; Rocheleau, 1994). The nature of farmer participation can differ substantially within IPM programs identified as being participatory, and participation may involve people and institutions such as policy makers, marketing agents, and nongovernmental organizations in addition to farmers, scientists, and extension workers. The intended results of participation differ among organizations and individuals. Therefore a key issue for any IPM program is to decide on which participatory methods to apply in order to achieve the desired results (see Figure 1-2, page xvii).

With a focus on results, these methods must address the issue of how to extend the program to a broad audience. Particularly if the target group includes limited-resource farmers, that audience may be huge, in the millions in most developing countries. The sheer size of the group creates a dilemma. How can the needs of local farmers be met through participatory IPM, while, at the same time, lessons learned that are cost-effective be extended to the broader audience? Meeting local needs requires farmer participation, but if participation is too intensive (involves many interactions between the farmer and the technical person), public financial resources for IPM programs may be exhausted before very many farmers are reached. Compounding the problem is the relative weakness of public extension programs in developing countries.

The appropriate tradeoff between (a) highly-intensive participatory IPM programs that can empower a relatively small group of farmers to make very knowledgeable IPM decisions versus (b) less-intensive participatory IPM programs that might reach larger numbers of farmers but impart a potentially shallower understanding per farmer, depends on several factors. First, the funding source (local, national, international) determines in part the target group and the relative importance of technology and information (T&I) spillovers across regional or national boundaries. A program that is primarily locally funded hopes to achieve as much depth of knowledge as possible and may be less concerned about the spread of technology beyond the local area, while an internationally funded program expects broad communication of information. Second, the structure, strength, and linkage of the existing research and extension (R&E) institutions influence the speed with which IPM T&I can be generated and disseminated. The stronger the capacity and the linkage of the overall R&E system in a country, the greater the chance a pilot program will be duplicated beyond the immediate area. Third, the specific commodities and pest problems selected, and the major technologies involved, can significantly affect how easily the IPM program can be expanded or "scaled up". IPM programs for a major crop such as rice may spread more easily than a program for minor crops such as vegetables, and if the pest problem is amenable to an improved technology that involves a product such as an improved seed or a grafted seedling, the solution will likely spread more quickly than if it is only amenable to improved management.

Fourth, homogeneity in both the physical and the socio-economic environments can greatly influence the ability to scale up an IPM program. For example, within very short distances within the Andes in South America and within the Eastern African Highlands, micro-climates, farm sizes, gender roles in agriculture, and cultural traditions for working and communicating closely with one's neighbors differ greatly. Fifth, the cost and local applicability of a technology can determine whether farmers will adopt it. Technologies that are cost effective and are appropriate for the local circumstances have a higher probability of adoption. Sixth, the educational level of farmers can have a major bearing on the speed and level of adoption of an IPM strategy. In general, farmers with higher levels of formal education are more likely to adopt technologies faster and perhaps in a more sustainable manner. While these differences underscore the importance of participatory approaches, they also indicate a need for a deliberate, proactive T&I transfer strategy, because the farmer-to-farmer approach for spreading IPM technology may proceed too slowly.

Successful generation and cost-effective dissemination of IPM requires a participatory approach that recognizes the comparative advantage of various participants (farmers, national and international scientists, extension workers, policy makers, input suppliers) and the importance of the environment in its many dimensions (natural resource, social, cultural, economic, and institutional). Designing and implementing such a participatory approach is difficult anywhere, particularly where farms are small and their numbers are great. When farms are larger and fewer and environmental rules more restrictive, as in the United States, farmers have an incentive and capacity to directly seek out information from knowledgeable people in the public and private sectors, including scientists and other experts, thus facilitating the spread of IPM. The challenge of developing and facilitating the spread of IPM principles (globalizing IPM) in developing countries is greater, but no less important.

The chapters in this book describe an approach to implementing appropriate IPM programs in selected developing regions of the world. It focuses to a significant extent on how participatory IPM research must be integrated with the diffusion process if widespread dissemination of IPM T&I is to occur, just as successful farmer-participatory IPM extension programs can not divorce themselves from scientists and other players upstream in the scientific knowledge/regulatory/marketing chain. The purpose of this book is to describe an experience with a set of similarly focused IPM programs and to place those programs in the context of other IPM efforts underway around the world. In the process, we trust that some light will be shed and debate stimulated on workable approaches for the cost-effective generation and diffusion of IPM knowledge in the developing world.

Structure of the Book

Chapter 2 reviews briefly the general approaches currently being used for IPM research/diffusion around the world. The participatory integrated pest management (PIPM) approach utilized in the context of the IPM CRSP is described in detail, setting the stage for examples presented in subsequent chapters.

Section II of the book includes five chapters that discuss how the PIPM approach is applied in five very different parts of the world: Asia, Africa, Latin America, the Caribbean, and Eastern Europe. Chapter 3 presents Asian examples of how IPM practices for different insects, diseases, weeds, and nematodes are developed and integrated in an IPM strategy. It describes how pest problems and IPM strategies are identified and prioritized through stakeholder meetings, participatory appraisals, baseline surveys, and crop-pest monitoring. Furthermore, Chapter 3 summarizes the process for designing and testing PIPM tactics and strategies through laboratory, greenhouse, and on-farm experiments for onion and eggplant in the Philippines and for eggplant and cucurbit in Bangladesh. It discusses the role of interactive transnational linkages, how interdisciplinary analysis has been facilitated, and the integration of social and gender analyses in the IPM program.

Chapters 4, 5, and 6 address topics similar to those in chapter 3, but for other regions. Chapter 4 focuses on examples from Africa, both east and west. The eastern Africa case targets maize/bean and sorghum/groundnut/cowpea farming systems common to eastern and northern Uganda. Selected horticultural crops such as tomatoes and potatoes are also addressed in the Ugandan program. The development and deployment of IPM packages in these systems are discussed. The West Africa case focuses on IPM of sorghum/millet/cowpea farming systems as well as bean/hibiscus in a peri-urban production system in Mali. The significant differences between the Africa and Asia cases illustrate the need to tailor participatory approaches to accommodate differences in natural resource allocations, culture, and institutions. Chapter 5 focuses on Latin America and emphasizes IPM strategies on non-traditional agricultural export (NTAE) crops such as snow-pea and broccoli IPM systems in Guatemala and potato and plantain systems in Ecuador. These commercial export commodities in Guatemala present a set of quality control issues and pre-export inspection protocols that need to be addressed in the context of an IPM strategy if the country is to succeed in developing a stable market for these products. Chapter 6 provides examples of participatory IPM protocols for sweet potatoes, hot peppers, and vegetable amaranth in the Caribbean, developed under the guidance of a regional research organization (Caribbean Agricultural Research Development, CARDI) given the small size of the individual countries. Chapter 7 considers an IPM program in a country, Albania, undergoing major economic and political transition.

The chapters in Section III focus on issues related to deploying IPM packages. Chapter 8 draws on examples from the various regions to examine the roles of public extension, cooperatives, non-governmental organizations (NGOs), and other private sector groups in the transfer and adoption of IPM. The roles of farmer field schools, simple messages, and other approaches to facilitating the spread of IPM knowledge are considered, with an eye toward the comparative advantage of various methods in different situations. Knowledge diffusion within a country is considered as well as diffusion across national and regional boundaries. The roles of international organizations are also considered. Methods for and examples of assessing adoption are discussed, with special attention to how social and gender issues are incorporated. Because many crop protection systems in developing countries are thin in scientific research and extension personnel, training programs that involve training outside the country, inside the country, and combinations of these are described and assessed.

The farmer field school (FFS) approach to disseminating IPM, discussed briefly in Chapter 8, is elaborated in more detail in Chapter 9, with examples from around the world. The FFS process, countries and institutions with major FFS programs, and the role of FFS in educating farmers and trainers are described. Evidence on FFS cost-effectiveness is discussed, as is the need for developing-country institutions to be cautious in their interactions with chemical companies when implementing a field-school approach.

The adoption of IPM strategies can be heavily influenced by policy or institutional issues. Chapter 10 provides methods for and examples of pesticide and IPM policy analysis in the Philippines, Uganda, Ecuador, and Jamaica. Because pesticides are imported by most developing countries, direct and indirect (e.g., exchange rate) trade policies can be as or more important than domestic policies in creating economic incentives for adopting IPM practices. Environmental regulations and their enforcement also critically affect pesticide use and hence the implementation of IPM practices. In many countries, agricultural commodities receiving particularly heavy doses of pesticides are export crops. Chapter 11 presents a detailed example of a pre-inspection IPM program for export crops in Guatemala. Drawing on a snow-pea example, the steps followed in institutionalizing an IPM quality control process in Guatemala are described.

Evaluating strategic IPM packages is the focus of Section IV. In Chapter 12, methods are presented for evaluating economic and social impacts of IPM adoption. Examples of farm- and local-level assessments of economic impacts in the Philippines, Bangladesh, and Uganda are presented, with examples of aggregate or market-level assessments for these countries and Ecuador. The role of geographic information systems (GIS) in assessing the impacts is discussed with examples from Jamaica, Uganda, Bangladesh, and Ecuador. Gender and social impacts are also considered, with examples from Mali, Uganda, Guatemala, and the Philippines. In Chapter 13, methods are presented for evaluating how IPM practices affect health and the environment. Detailed examples are presented from impact assessments in the Philippines and Ecuador, with suggestions of practical, cost-effective assessment tools. In Chapter 14, the myriad ways in which gender affects the adoption and impacts of IPM are discussed, as women prove to be both more involved than expected in household decisions concerning pesticide use and less able than men to obtain access to IPM information and technologies. Even in regions where women are unlikely to view themselves as "farmers," women often influence pest management strategies through their roles in household budget management and decision-making regarding expenditure for pesticides and labor. Obtaining accurate information about women's roles in the work and decision processes that ultimately result in household pest management choices is a critical component of participatory IPM research.

Section V (Chapter 15) presents conclusions and lessons learned from the IPM experiences in Asia, Latin America, the Caribbean, Africa, and Eastern Europe. Cost-effective globalization of IPM is possible but will demand a retreat from the one-size-fits-all approaches of the past, and a recognition that more attention must be devoted to effective collaboration and cooperation among international, regional, and national organizations, and between these groups and farmers and consumers.

References

Bentley, Jeffrey W., Gonzalo Rodríguez, and Ana González. 1994. Science and people: Honduran campesinos and natural pest control inventions. *Agriculture and Human Values*, Winter: 178-182.

Dlott, Jeff W., Miguel A. Altieri, and Mas Masumoto. 1994. Exploring the theory and practice of participatory research in U.S. sustainable agriculture: A case study in insect pest management. *Agriculture and Human Values*, Spring-Summer: 126-139.

Food and Agricultural Organization. 1967. Report of the first session of the FAO Panel of Experts on Integrated Pest Control, Rome.

Norton, George W., Edwin G. Rajotte, and Victor Gapud. 1999. Participatory research in Integrated Pest Management: Lessons from the IPM CRSP. *Agriculture and Human Values* 16: 431-439.

Rocheleau, Dianne E. 1994. Participatory research and the race to save the planet: Questions, critique, and lessons from the field. *Agriculture and Human Values*, Spring-Summer: 4-25.

Röling, N., and E. van de Fliert. 1994. Transforming Extension for sustainable agriculture: The case of Integrated Pest Management in rice in Indonesia. *Agriculture and Human Values*, Spring-Summer: 96-108.

Thrupp, L.A., and R. Haynes. 1994. Special issue on participation and empowerment in sustainable rural development. *Agriculture and Human Values*, Spring-Summer: 1-3.

Van de Fliert, E. 1993. Integrated Pest Management: Farmer field schools generate sustainable practices, a case study in Central Java evaluating IPM training. Wageningen, The Netherlands: Agricultural University of Wageningen, Thesis 93-3.

— 2 —

Participatory Integrated Pest Management (PIPM) Process

George W. Norton, Edwin G. Rajotte, and Gregory C. Luther

Introduction

Successful integrated pest management programs require interactions among scientists, public and private extension, farmers, policymakers, and other stakeholders. Defining the appropriate nature of those interactions is difficult because research and dissemination activities require financial and human resources, and because farmers, scientists, and extension workers have comparative advantages in specific aspects of the knowledge generation and diffusion process. In addition, the ease of transferring technologies depends on the environmental sensitivity of the technologies and on environmental, cultural, and other sources of diversity within countries.

With limited resources, scientists or extension workers cannot interact directly with all farmers. Therefore it is essential for farmers to generate many of their own IPM technologies and to learn from each other, and for IPM knowledge to diffuse through a variety of channels. Farmers know a lot but not everything about their pest problems. They often incorrectly diagnose their problems because many pests are difficult to see (Bentley et al., 1994). Therefore interactions among farmers, researchers, and extension workers are needed to help identify the principal causes of and potential solutions to pest damage observed by farmers. The question is how to obtain those interactions in a sustainable, cost-effective manner.

Participatory Approaches to IPM

Farmer-participatory approaches have been used for many years in extension systems relying on participatory group-learning methods, but participatory IPM programs took a significant step forward in the late 1980s with the initiation of the "Farmer-Field-School" (FFS) approach to

Figure 2-1. Grower meeting to evaluate IPM training program in Mali.

rice IPM in Indonesia (Kenmore, 1991). A summary of this approach can be found in Roling and van de Fliert (1994), and in Chapter 9 below; in brief, FFS stresses the importance of farmers growing a healthy crop, observing their fields weekly, conserving natural enemies, experimenting themselves, and using relevant, science-based knowledge.

A farmer-training program is held with groups of about 25 farmers (often broken into sub-groups) (Figure 2-1). The "field schools" last for an entire growing season in order to take the farmers through all stages of crop development. Little lecturing is done, with farmers' observations and analyses in the field providing key components of the process. Farmers and trainers discuss IPM philosophy and agro-ecology, and farmers share and generate their own knowledge (van de Fliert, 1993; Yudelman et al., 1998).

Applied first in Asia with support from FAO and USAID, FFSs spread to Africa and Latin America and now are used in vegetables and many other crops as well as rice. FFSs are practiced by national agricultural research systems as well as by NGOs. For example, the Philippine government uses FFSs as its principal approach to IPM diffusion, while CARE relies on them in a major way in Bangladesh and several other countries. The model has been well received and is, where resources permit, a viable approach for achieving in-depth farmer knowledge of sustainable pest management practices.

Despite the success of targeted FFS participatory IPM programs, overall spread of IPM in developing countries remains relatively limited, and the need has been recognized to extend participatory IPM research and extension programs in other dimensions if widespread adoption is to be achieved. There is a need to improve the participatory nature and scientific rigor of IPM research conducted by scientists in agricultural research systems, and to define more clearly the linkages between farmer-based research and extension programs such as the FFSs and the general supply of national and international scientific knowledge. The initial FFS programs in rice had the advantage of building on a wealth of scientific information already widely available for rice. IPM programs for vegetables and many other crops have not had this luxury, and perhaps for this reason have proceeded more slowly. An optimal IPM program has strong linkages between farmers and upstream (fundamental and applied) research, and between farmers and market information as well.

A second critical issue is how an IPM program can cost-effectively diffuse information to millions of small farmers. FFS programs have helped thousands of farmers to understand IPM and to gain the skills needed to develop methods for managing their pests. However, they cost roughly $40 to $50 per farmer trained in the Philippines, Indonesia, and Bangladesh (Quizon, Feder, and Mugai, 2000; Tim Robertson, CARE-Bangladesh personal communication, 1999). Unless farmers who receive FFS training impart their knowledge to large numbers of their neighbors and retain the IPM knowledge for a long time and use it on more than one crop, relatively few farmers can be reached before training budgets are exhausted. Recent studies sponsored by the World Bank examined the issues of IPM knowledge retention and diffusion (Quizon, Feder, and Mugai, 2000; Rola, Jamias, and Quizon, 2002). Results were not especially encouraging. They found significant retention but little diffusion. Impacts on yield and pesticide use have also been questioned (Feder, Murgai, and Quizon, 2004).

The important questions are how to maximize IPM information generated and spread for the dollars and effort spent, and how to obtain a reasonably large amount of knowledge diffusion given tight public budgets. The answers are likely to vary by country and to require a multiple-pronged approach that considers the nature of the crop, the pests, the technologies, the need to involve public and private sectors, the types of farmers, the quality of existing research and extension organizations, and other factors. The purpose of this chapter is to describe, in general terms, a participatory IPM approach that is currently being tested in several countries, with the details of specific applications to be provided in subsequent chapters. The approach includes a general set of principles and a flexible participatory IPM (PIPM) process that continues to evolve and be tailored, to some extent, to each specific country setting.

PIPM Principles

The guiding principles of the PIPM process include the following:

1) Farmers and scientists both participate so that they can learn from each other;
2) The IPM research program is multidisciplinary and includes social scientists as well as biological scientists (fundamental and applied);
3) Participation extends to the output and input marketing sectors as well as regulatory and other government institutions;
4) Appropriate linkages are made to sources of knowledge external to the country;
5) Diffusion of IPM knowledge to producers involves all relevant channels in the public and private sectors;
6) Multiple methods for IPM diffusion are utilized as appropriate, given the nature of the IPM information, time and resources available, and characteristics of recipients;
7) IPM research is institutionalized in existing organizations, where possible, in a way that it is sustainable over time without extraordinary external resources;
8) All activities are subject to impact assessment.

The goal of PIPM is to increase incomes for the whole population while reducing health and environmental risks associated with pest management. Achieving this goal requires good science, farmer involvement, and recognition of the myriad factors that influence farmer decision-making. It means recognizing that generating and spreading IPM knowledge requires scarce resources, and just reaching a few thousand farms, large or small, achieves little. It is easy to get absorbed in arguments over details of specific participatory methods or research and extension techniques and lose sight of the goal. The reality is that no single research or extension method works everywhere or even for every pest problem in a specific site. Therefore the PIPM process described below is a broad approach designed to be flexible, within the bounds of the principles described above.

The PIPM Process

IPM is a process that:

1) builds on fundamental information about the pests and their environment,
2) identifies solutions to pest problems for specific crops (or livestock, lawns, schools), and
3) facilitates the spread of IPM strategies.

IPM accomplishes these tasks by involving the appropriate people at each stage in the process so there is no disconnect among the stages. Therefore it begins by identifying the relevant collaborators and other stakeholders (Figure 2-2). These people help determine possible sites for experimental work, specific commodity foci, and other aspects that define the broad parameters of the program focus. The initial group of collaborators and stakeholders may include scientists, public and private extension workers including representatives of non-governmental organizations (NGOs), farmers, representatives from farm organizations, local leaders, public officials, and others.

Site selection is critical since on-farm experimental work requires representative areas and locations where the logistics allow scientists to regularly visit the field. Such sites are reasonably close to an experiment station for the initial primary sites. Secondary sites can be selected in major growing areas to facilitate additional on-farm testing and obtain feedback from farmers on IPM strategies developed on the more intensive on-farm experimental sites, but where scientists are not present as often during the season. If the IPM program has international involvement, it may be important to select a research site that is representative of a major agroecosystem in a multi-country region.

Scientific collaborators are chosen to represent an appropriate disciplinary mix, and typically might include expertise in entomology, plant pathology, weed science, nematology, economics, and sociology/gender analysis.[1] Farmer collaborators are chosen based on interest in working with scientists on their farms. Collaboration with farmers who are members of a cooperative or other group can work well, because of the additional feedback that may be obtained from neighboring farmers, and the increased capacity to spread information. Other stakeholders often include consum-

[1] The IPM CRSP, being an international collaborative research program, included these disciplines in both its foreign and domestic scientists.

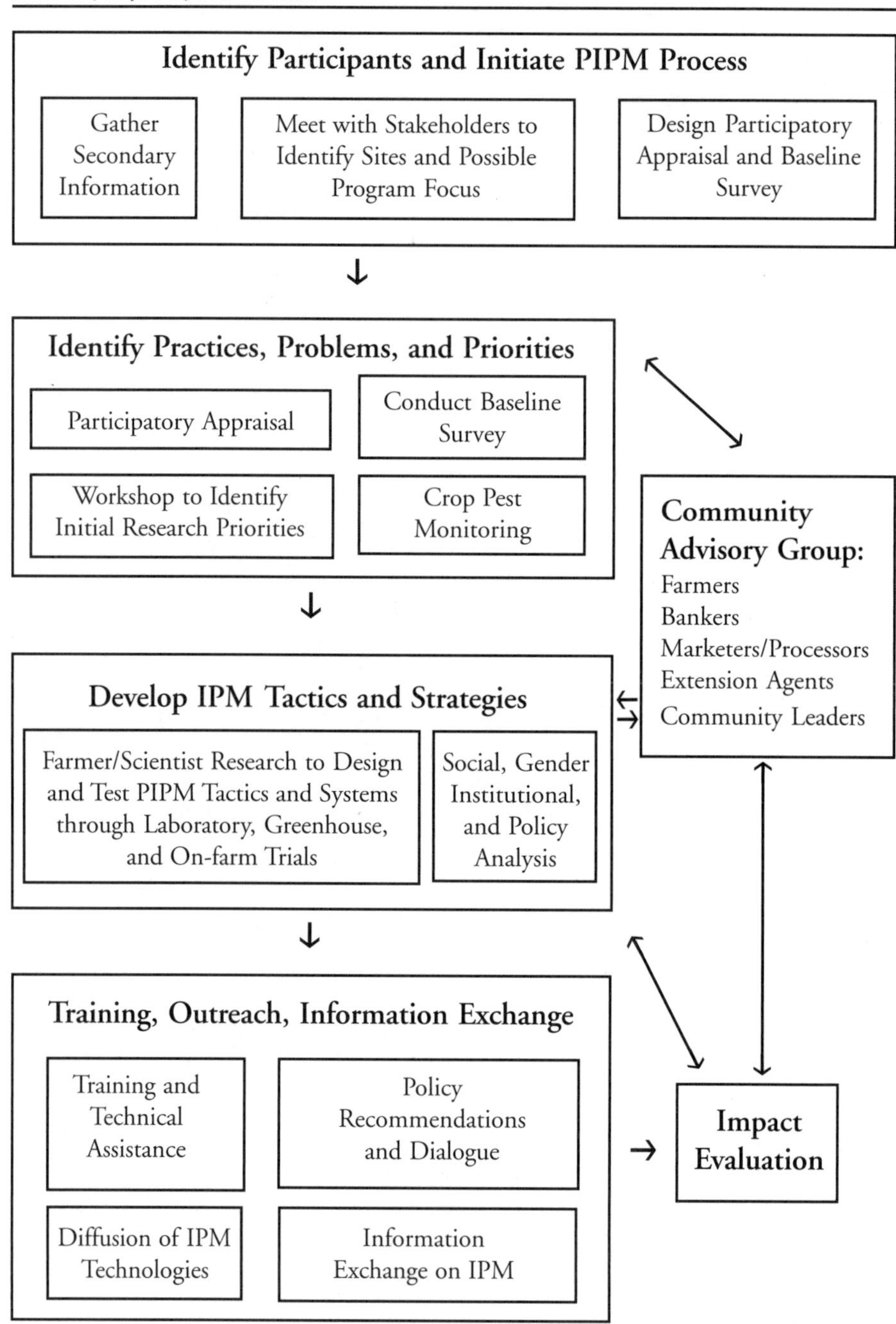

Figure 2-2. Suggested PIPM process.

ers, government officials, NGO representatives, marketing agents, and lenders.

Gathering Fundamental Information

Several fundamental information-gathering activities are conducted in a PIPM program to help establish research priorities. Primary among these activities are gathering secondary information, conducting a baseline survey and participatory appraisals, and monitoring the abundance of pests and beneficial organisms in the crop system. *Secondary information* is collected on the production and value of specific crops locally, nationally, and regionally; on the magnitude of pest losses due to specific pests, on the amount of previous and existing IPM research activities in the public and private sectors, and on important socioeconomic factors that may influence an IPM program.

The baseline survey may include regions beyond the local research sites in areas where the same crop/pest complex is known to be prevalent. A survey of 300-400 farmers can serve to identify farmers' pest perceptions, pest management practices,and decision-making processes; basic socioeconomic characteristics; and other information. It provides a baseline against which evaluation can occur down the road.

A participatory appraisal (PA), which uses a less-structured information-gathering technique than a survey, complements the survey because it allows follow-up on issues that need clarifying (Figure 2-3). The PA process begins with a brief training session for scientists on PA methods, just prior to going into the field. The PA takes one to two weeks for a typical IPM site and helps to foster interdisciplinary relationships among collaborators as well as to assist with research priority-setting. Additional details on the PA process are provided in the next four chapters and in publications such as Litsinger et al. (1995).

A fourth key information-gathering activity is *basic identification* and *field monitoring of pests and beneficial organisms* for at least two years. Precise identification of pests is crucial if subsequent IPM strategies are to build on knowledge from other locations or sources, because pest species or strains can behave differently and there is little use in duplicating what is already known (Irwin, 1999). Pest monitoring is critical because pest abundance and timing must be assessed to estimate the economic significance of the pests, and to develop subsequent solutions to the problems. Farmer-collaborators selected for on-farm experiments can assist with the monitoring.

Figure 2-3. Interviewing farmers during a participatory appraisal in the Philippines.

Developing Solutions to Pest Problems

While crop/pest monitoring is underway, farmers and scientists work together to *design, test, and evaluate IPM tactics and management strategies.* In many cases, pot, greenhouse, or micro-plot experiments are appropriate aids to assessing IPM tactics (practices) and, in virtually all cases, basic farm-level experiments of IPM tactics and strategies are conducted (Figure 2-3). At least two years of farm-level experiments are required before any particular practice can be recommended (see Figure 2-4, page xvii). Concurrently, *social science research* is undertaken on policies, institutions, and social factors that may influence the development and adoption of IPM. For example, the role that gender plays in decision-making and how benefits are distributed within the household may affect the success of an IPM program. Establishment of a community advisory group may help provide feed-back on a continuing basis.

There is sometimes a debate about the value of research conducted by scientists versus farmers. This debate intensified with the introduction of farming-systems and farmer-participatory research paradigms in the 1970s and '80s. While the debate has helped focus attention on the importance of including clientele in the research process, the reality is that an effective IPM research program often involves farmers, scientists from national experiment stations, university scientists, the international scientific community, public extension workers, extension workers at NGOs, and others. It is not a question of farmer-led research versus experiment station research. The question is how to optimally involve all the players who might contribute to a successful IPM program. Each player has an appropriate role and the successful program recognizes how to sequence components and interactions among them. Scientists will achieve little success in solving pest management problems without involving farmers, and farmers, even working closely with extension workers and other technically trained people, will make limited progress without input from upstream scientists.

Spreading IPM Management Strategies

As research output is generated, outreach and information exchange beyond the locations of the on-farm experiments become important. Scientists interact with public extension, non-governmental organizations (NGOs), and private agribusiness firms and cooperatives. The relationship among these groups and their relative importance differ by country and farming system. FFSs are one way for extension and NGO groups to generate and spread IPM information when sufficient resources are available and in-depth understanding is required (See Chapter 9). For some technologies, simple messages can be cost-effectively spread through media campaigns, or through the private sector, especially if the messages are embedded in products. Increasingly, information flows by electronic means in addition to publications, workshops, field days, and other traditional methods. A community advisory group may be helpful in spreading IPM information locally. However, advisory councils require empowerment by scientists if they are to be effective.

Because most IPM research is conducted on a relatively small number of farms where logistics allow interactions with scientists, it is important to test promising IPM strategies over a broader area than that where the research is focused. Simple experiments, demonstration plots, and field schools with farmer-led research can be strategically located around the country for the purpose of providing feedback to scientists about the need to adjust technologies to specific local conditions. These activities can test packages or integrated strategies and need not be conducted under the strict statistical designs that are necessary at earlier stages in the research process.

The role of the private sector can be important, particularly when IPM research results are embedded in seeds or other products such as resistant varieties and grafted seedlings. In more developed countries, crop consultants can impart information on cultural or management practices as well, although private impartial consultants (as opposed to salespersons) are yet to make many inroads among limited-resource farmers in developing countries, except where there are strong linkages to the market through a cooperative or export firm that hires technical people to work with farmers.

The international centers play a critical part, both in terms of scientist training and in extending and testing research in countries beyond where the initial research is conducted. The existence of rice, vegetable, potato, and wheat networks coordinated by international centers provides a mechanism for transferring information and for bringing it back to the scientists.

Impact assessment is also essential to the PIPM process. Assessing the economic, health, environmental, and social implications of alternative IPM technologies feeds directly into recommendations for farmers and policy makers. Aggregate assessments can also help in setting priorities and in justifying an IPM program to funding sources. Aggregate assessment requires adoption analyses, which may be *ex ante* (before the research) and include expert opinion, or *ex post* (after the research) and include data on actual adoption. Adoption analysis can be useful for assessing issues that might be addressed to increase adoption.

Successful IPM programs are institutionalized in national research systems. In many cases, institutionalization will involve short- and long-term training for researchers and extension workers.[2] One of the most cost-effective and least disruptive types of graduate level training involves a "sandwich" program in which the student takes classes and receives the degree from a developing-country university, but spends a semester or two taking courses and conducting research at a U.S. university or international agricultural research center (IARC).

Institutionalization of IPM research also requires cost-effective programs that mesh with other research programs. Particularly for disciplines such as economics and plant breeding, IPM may be only one component in a broader research program. Therefore, IPM planning and review can be a piece of an overall research review and planning process. External resources may supplement the program in its early stages, but sustaining the program over time requires a commitment to make PIPM a part of the recurring research program.

[2] The IPM CRSP has met that need through a combination of short-term (2 weeks to 6 months) training at U.S. universities and international centers, and graduate education with degrees from either U.S. or regional universities in Africa, Asia, and Latin America.

The PIPM principles and process presented very briefly in this chapter are elaborated on through examples in the remaining chapters. One key to the process is broad yet cost-effective participation.

References

Bentley, Jeffrey W., Gonzalo Rodríguez, and Ana González. 1994. Science and people: Honduran campesinos and natural pest control inventions. *Agriculture and Human Values*, Winter: 178-182.

Feder, G., R. Mugai, and J.B. Quizon. 2004. Sending farmers back to school: the impact of farmer field schools in Indonesia. *Review of Agricultural Economics* 26 (Spring): 45-62.

Irwin, Michael E. 1999. Implications of movement in developing and deploying Integrated Pest Management strategies. *Agricultural and Forest Meteorology* 97: 235-248.

Kenmore, P.E. 1991. Indonesia's Integrated Pest Management, a model for Asia, Manila, Philippines. FAO Intercountry IPC Rice Program.

Litsinger, James, George W. Norton, and Victor Gapud. 1995. Participatory appraisal for IPM research planning in the Philippines. IPM CRSP Working Paper 95-1, Virginia Tech, September.

Norton, G.W., E.G. Rajotte, and V. Gapud. 1999. Participatory research in Integrated Pest Management: Lessons from the IPM CRSP. *Agriculture and Human Values* 16: 431-439.

Quizon, J., G. Feder, and R. Murgai. 2001. A note on the sustainability of the farmer field school approach to agricultural Extension. *Journal of International Agriculture and Extension Education* 8 (Spring): 13-24.

Rola, A., S. Jamias, and J. Quizon. 2002. Do farmer field school graduates retain and share what they learn? An investigation in Ilailo, Philippines. *Journal of International Agriculture and Extension Education* 9 (Spring): 65-76.

Röling, N., and E. van de Fliert. 1994. Transforming Extension for sustainable agriculture: The case of Integrated Pest Management in rice in Indonesia. *Agriculture and Human Values*, Spring-Summer: 96-108.

Van de Fliert, E. 1993. Integrated Pest Management: Farmer field schools generate sustainable practices, a case study in Central Java evaluating IPM training. Wageningen, The Netherlands: Agricultural University of Wageningen, Thesis 93-3.

Yudelman, M., A. Ratta, and D. Nygaard. 1998. Pest management and food production: Looking to the future. Food, Agriculture, and the Environment Discussion Paper 25, International Food Policy Research Institute, Washington, D.C.

— II —
Developing Strategic IPM Packages

– 3 –
Developing IPM Packages in Asia

Sally A. Miller, A.M.N. Rezaul Karim, Aurora M. Baltazar,
Edwin G. Rajotte, and George W. Norton

Introduction

IPM programs have grown rapidly in Asia over the past 20 years, stimulated in part by donor-supported projects in Indonesia, the Philippines, Bangladesh, Vietnam, and other countries, and by development of significant IPM programs in national agricultural research systems (NARS) in India and elsewhere. One of the largest IPM efforts initially focused on rice with the assistance of the International Rice Research Institute (IRRI) in developing programs with NARS throughout the region, and the Farmer Field School (FFS) programs developed with support from the U.S. Agency for International Development (USAID) and the Food and Agricultural Organization (FAO). The FFS programs were implemented through both governmental and non-governmental organizations (NGOs).

Attention to vegetable crops was stimulated in part by the efforts of the Asian Vegetable Research and Development Center (AVRDC), which supported IPM programs in countries throughout the region. IPM CRSP activities in Asia began with the establishment of the Southeast Asia site in the Philippines in 1994, and the South Asia site in Bangladesh in 1998. AVRDC and IRRI are partners in the IPM CRSP program together with the host country and U.S. institutions. Donors from the U.K. and elsewhere established targeted IPM research programs in specific countries on a variety of additional crops. Outreach efforts in vegetable IPM have been facilitated by FFS programs run by CARE, especially in Indonesia and Bangladesh.

Research at the two IPM CRSP sites has centered on IPM for vegetables produced intensively or in rotation with rice. There are similarities between the sites in cropping systems, pest problems, and socioeconomic issues, and many of the U.S. and other international team members have worked at both sites. While rice IPM has been firmly established in Asian countries for several years (Shepard, 1990; Ooi and Waage, 1994; Savary et al. 1994; Teng, 1994), investments in IPM research, development, and information dissemination in vegetable cropping systems have historically been lower (Guan-Soon, 1990). As vegetable supply and demand has grown over the past few years, vegetable production has become an important source of income for many resource-limited farmers. Unfortunately, vegetables are also subject to damage by numerous insect pests, weeds, and diseases. Misuse of pesticides on vegetables is common, resulting in increasingly severe pest problems, and increased incidence of environmental contamination and pesticide poisoning; Tejada et al., 1995; Tjornhom et al., 1997; Lucas et al., 1999; Rahman, 2003; Rashid et al., 2003).

Identifying and Prioritizing IPM Problems and Systems

Vegetable IPM programs under the IPM CRSP began in both the Philippines and Bangladesh with participatory appraisals (PAs), stakeholder meetings, and baseline surveys, followed by several seasons of crop-pest monitoring to determine and prioritize pest problems and socioeconomic questions, and to assess constraints and opportunities.

The Philippines

In the Philippines, an initial meeting was held in March 1994 at the Philippines Rice Research Institute (PhilRice) in Muñoz, Nueva Ecija, with participants from PhilRice, the University of the Philippines at Los Baños (UPLB), the International Rice Research Institute (IRRI), the Asian Vegetable Research and Development Center (AVRDC), FAO, USAID, Virginia Tech, and Pennsylvania State University (Gapud et al., 2000). These stakeholders established the participatory process that formed the basis of IPM research, development, and outreach activities of the site. Other IPM programs in the Philippines were reviewed, crops were prioritized, and potential research sites were determined (Litsinger et al., 1994). PhilRice, an agency of the Philippines Department of Agriculture with a strong national research and development system, was designated the lead national institution for IPM CRSP activities. Other stakeholders were identified, including farmers and their families, a local cooperative, city agriculturalists, Landbank, barangay (village) captains and councils, policy makers in the Philippines Department of Agriculture, the National Onion Growers Cooperative Marketing Association (NOGROCOMA), chemical and seed companies, and the Philippine Fertilizer and Pesticide authority.

San José, Nueva Ecija, an area in the Central Luzon approximately 130 km north of Manila, was selected as the initial site for on-farm experiments. Local agricultural officials were consulted in the selection of nine barangays (neighborhoods) in or near San José and the nearby towns of Santo Domingo and Talavera. Farmers in these towns were interviewed by the site selection team regarding their pest problems, crops, and constraints particularly, but not exclusively, related to pest management. Finally, in consultation with municipal agriculturalists and technicians, six barangays (Santo Tomas, Abar 1st, Kita-kita, Sibut, Manicla, and Palestina) in San José were selected.

A baseline survey was conducted in June 1994 with 300 farmers to determine their perceptions of diseases, weeds, insect pests and their natural enemies, and to assess farm characteristics and pest management practices. Some pertinent results of the survey were: 1) onion was the predominant vegetable crop planted after rice; 2) 77% of the farmers surveyed planted vegetables after rice, 18% planted only rice, and 5% planted only vegetables; and 3) pesticides were heavily used on some farms, with up to 10 applications on rice and 24 applications per vegetable crop.

These data were used to plan the PA, which was carried out over a two-week period in July 1994. PA procedures were discussed among the participating scientists in a brief workshop prior to going into the field. During the PA, farmers from six villages were interviewed, and additional interviews were conducted at local cooperatives, the Landbank, and a local hospital (to determine incidence of pesticide poisonings); and with the city agriculturalist, agricultural extension workers, barangay captains and counselors, pesticide dealers, assistant secretaries in the Department of Agriculture, and representatives of the National Economic Development Authority, AVC Chemical Corporation, USAID, NOGROCOMA, and the Philippine Institute of Development Studies.

Scientists who participated in the PA visited with farmers and other stakeholders in small teams during the mornings and spent the afternoon debriefing, analyzing relationships, formulating follow-on questions, and planning additional PA activities the for next day.

Litsinger et al. (1994) present results of the PA in detail. In brief, average farm size ranged from 0.9 ha (Abar 1st) to 2.4 ha (Santo Tomas). Five dominant cropping patterns were observed: 1) rice-rice, 2) rice-fallow, 3) rice-vegetables, 4) rice-vegetables-vegetables, and 5) vegetables-vegetables. A wide range of vegetables was planted after rice, the most common being members of the onion family (onion, garlic, shallots), followed by eggplant, string beans, mungbeans, sweet and chili peppers, various squashes, tomato, green corn, and Chinese cabbage, known locally as pechay. Farmers perceived the stem borer and "worms" to be the most important pests of rice, although the golden snail, rats, "weeds", numerous other insect pests, and rice tungro virus were also mentioned by at least 10% of the farmers. Thrips, armyworms, weeds, bulb rot, and damping-off were mentioned as important pests of onions by 34% or more of the farmers surveyed. Although the sample size of farmers responding to questions about the remaining crops was small, several pest problems emerged as most important, including fruit and shoot borer of eggplant and "worms" of peppers, string beans, and bitter gourd.

Insecticide use on rice was higher than recommended by IRRI (1995), averaging 1.5 applications in the seedbed and 2.3 to the main crop. Seventy percent of the applications were calendar-based. Both banned and unsafe insecticides were used, mainly applied using a lever-operated knapsack sprayer that generally results in considerable pesticide contamination to its operator. The use of synthetic pyrethroids, which have a negative impact on natural enemies, was also mentioned by a number of farmers. Weeds were managed by a combination of cultural, mechanical, and chemical methods. Disease control practices were not mentioned. High amounts of insecticides were applied to onions, with an average of 7.5 applications per crop, 87% of which were calendar-based. For weed control, farmers applied a rice straw mulch to preserve soil moisture and suppress weeds. Hand weeding, herbicide use, rice hull burning, and plowing and harrowing were also practiced. Various fungicides and even water were applied to control bulb rot. For eggplant, farmers attempted to control the fruit and shoot borer with 30-50 insecticide applications during the time of fruiting, without great success. Relatively high numbers of applications of insecticides were also reported for the remaining vegetable crops. Socioeconomic factors affecting rice production included lack of reasonable credit, high cost of irrigation, and low price at harvest. For onions, socioeconomic factors were similar but also included expensive seed, high input costs, price fixing by traders, lack of storage facilities, and limited technical support.

A large majority of the farms were operated by nuclear families. Land preparation, irrigating, and spraying were the exclusive responsibility of male family members. Women, who often decided what and when pesticides were to be applied, and also purchased them, primarily did the weeding. However, marketing of pesticides was often directed toward men. Men and women shared harvesting and marketing responsibilities.

Statistics were also gathered on marketing, credit, and land tenure. Onion marketing is complex in San José because San José is a center for onion production. Onions are produced for consumption in the Philippines and for export to other countries in Asia, primarily Japan, Singapore, and Malaysia. The large-bulbed Yellow Granex onion was produced for export. Farmers sold onions directly to traders or joined a cooperative for storage and marketing of onions, and for obtaining credit. Credit came from a variety of sources, and interest rates varied dramatically. Sources of credit included personal loans from friends, local moneylenders, traders, and banks such as the Land Bank. Most farmers in Nueva Ecija were shareholders, and various crop-sharing arrangements were made with the landowners. The barangay of Palestina, in which 70% of the farmers were owners, was an exception.

A research planning workshop held at the end of the process was another element of the PA. Crops and pests were prioritized, working groups prepared specific plans for whole-group discussion, and budgets were prepared. The group decided to focus on onion, because it is the principal vegetable crop of the area; eggplant, which is important in the Filipino diet and is also the recipient of a large amount of insecticide; and string bean, due to its pest profile, high degree of pesticide use, and crop importance. Questions and ideas for pest management and socioeconomic research were listed and prioritized during brainstorming sessions. Key examples of focal points for research and eventual dissemination of results are described in subsequent sections of this chapter.

During the first two years of the IPM CRSP in the Philippines, scientists worked closely with farmers to identify and prioritize pests on the targeted vegetable crops. Although farmers identified a number of specific problems during the PA, the general terms "weeds" and "worms" were often reported. Diseases were rarely mentioned, which is not surprising as many diseases are difficult to diagnose in the field and symptoms may be attributed to general problems with fertility, water management, etc. During the course of pest monitoring and surveys, researchers discovered that the rice root knot nematode, *Meloidogyne graminicola*, was the cause of serious yield losses of onions (Gergon et al., 2001a). Farmers had previously believed that the symptoms observed were due to insect pests or air pollution. In addition to the root knot nematode, the following pests were identified as most important and served as the focus for further research: common cutworm of onion (*Spodoptera litura*), eggplant fruit and shoot borer (*Leucinodes orbinales*) (see Figure 3-1, page xviii), green leafhopper of eggplant (*Amrasca biguttula*), and the weeds purple nutsedge (*Cyperus rotundus*), *Cleome viscosa*, *Phyllanthus amarus*, and *Trianthema portulacastrum*.

Bangladesh

In Bangladesh the PA was conducted over a ten-day period in August 1998 by scientists from the Bangladesh Agricultural Research Institute (BARI), the Bangladesh Rice Research Institute (BRRI), AVRDC-Bangladesh, CARE-Bangladesh, Institute for Post Graduate Studies in Agriculture (IPSA), AVRDC, IRRI, Virginia Tech, Penn State, Purdue, and UPLB. BARI, located in Joydebpur, was designated the lead national institution for the Bangladesh site. The principal site chosen for on-farm research was Kashimpur, located in the Gazipur district, 16 km from Joydebpur, but the PA included activities in Comilla and Sripur as well.

As in the Philippines, the PA began with a brief workshop on participatory appraisal methods and on the state of the vegetable sector and IPM in rice-vegetable systems in the country. Scientists and NGO representatives from the institutions listed above were broken into four groups, with field visits in the mornings and debriefings in the afternoons. Information was collected and discussed on seasonal cropping patterns, major pests by crops, pest management practices, information sources, markets, credit sources, land tenure, production constraints, economic factors, and gender/social/family issues that might impinge on IPM. Finally, a two-day workshop was held to prepare preliminary research plans for the program.

It was clear from the PA that eggplant, tomato, cabbage, a variety of gourds, cucurbits, okra, onion, and country beans were especially important among a large number of vegetables in need of IPM programs. Heavy insecticide and fungicide use was noted, but little herbicide use. Export markets were small, but with some potential for growth. Many of the same pests found in the Philippines were also a problem in Bangladesh, for example fruit and shoot borer and bacterial wilt on eggplant.

A baseline survey was also designed and conducted in a manner similar to that described above for the Philippines (Hossain et al., 1999). The survey was conducted in two villages (Enayetpur and Barenda-Noyapara) in Kashimpur Union and two (Aahaki and Joyertek) in Konabari Union, areas of highland to medium highland suitable for rice-vegetable cultivation. Three hundred male farmers, operating small, medium and large farms, and 100 female farmers from the same four villages, were selected. Female enumerators interviewed female farmers. The farmers were predominantly middle-aged, and the majority of males were illiterate. In contrast, all of the females were literate. About one-fourth of the males and none of the females had received training in rice and vegetable production. Farm size was less than one ha on average. Although both rice and vegetables were produced, the area is considered an intensive vegetable production area. A wide variety of vegetable crops was produced, depending on the season. The major cropping patterns were: vegetable-vegetable-vegetable (23%), fallow-fallow-rice (14%), vegetable-vegetable-rice (9%), vegetable-rice-vegetable (8%), fallow-vegetable-vegetable (7%), vegetable-fallow-vegetable (7%), fallow-rice-vegetable (6%), and vegetable-green manure-vegetable (6%). The major insect pests reported by farmers were caterpillars, aphids, ants, and fruit fly; diseases included mosaic virus, leaf blight, stem rot, and leaf spot. The majority of farmers reported applying insecticides 2-4 times per vegetable crop. Vegetables were sold to traders or to local markets. The use of credit was low; 16% of male farmers in Barenda-Noyapara village borrowed money from friends, neighbors, moneylenders, or a cooperative. However, 32% of female farmers reported borrowing money from friends and neighbors.

During the first year of crop and pest monitoring, several insect pests, weeds, and diseases emerged as significant problems in the evaluation area (Islam et al., 1999). Tomato yellow leaf curl disease (geminivirus) was the most serious problem of tomatoes identified, while late blight and early blight were also considered important in tomato. Mosaic virus in bottle gourd, root rot of cauliflower and cabbage, leaf rot of Chinese cabbage, bacterial wilt of eggplant, root knot nematode in country bean and bunching onion, and leaf blight of bunching onion were also reported.

Insect pests were monitored on eggplant, tomato, cabbage, okra, yard-long bean, and cucurbit crops in the Kashimpur area. Fruit and shoot borer was the main pest on eggplant, damaging about 16% of fruits. Other pests of moderate importance were jassids (leafhoppers), aphids, and white fly. On tomato, aphids were the dominant pest, infesting 29-73% of the plants; other pests were white fly and mite. Cabbage was mainly damaged by *Spodoptera* caterpillars. Pest infestation on okra and yard-long bean was low. Low infestations of noctuid caterpillars on okra and fruit fly on yard-long bean were observed. Cucurbit fruit fly was the main pest on all kinds of cucurbit crops in Kashimpur area; average infestation rates were 44% in white gourd, 32% in snake gourd, 25% in ribbed gourd, and 29% in bottle gourd.

Vegetable fields were generally hand-weeded. Dominant weeds were *Cynodon dactylon* in cauliflower and bottle gourd, *Cyperus rotundus* in eggplant, and *Eleusine indica* in cabbage, tomato, and radish.

The PA and baseline survey identified a host of potential areas that could benefit from the participatory IPM research approach. Research activities were focused on crop rotations that included highland eggplant (brinjal) and gourds and medium land cauliflower, gourds/cucumbers, tomatoes, and onions. Topics included varietal screening for resistance and grafting to manage bacterial wilt; root knot nematode, and fruit and shoot borer in eggplant; poison bait/pheromone traps for fruit fly on pumpkin gourd; cultural practices for weed management; and assessments of crop loss due to weeds. Socioeconomic topics were assessment of impacts of IPM systems and technologies, analysis of factors affecting IPM adoption, role of women in vegetable production and IPM decision making, and understanding the pricing and marketing context for the principal vegetable crops (IPM CRSP, 1999).

Designing and Testing Participatory IPM (PIPM) Tactics and Systems through Laboratory, Greenhouse, and On-farm Experiments

The multidisciplinary process by which PIPM tactics and systems are conceived and evaluated involves scientists and technical staff from host-country institutions, as well as scientists from cooperating institutions outside the host country. The process is the same in both sites in Asia, and some experiments are similar at both sites due to some similarity of crops and pest-management issues. Farmers and other stakeholders are consulted in the process as a means of ensuring that the most important pest management issues are addressed. Farmers may already employ specific pest-management practices that can be appropriate for wider application beyond their own farm or village once they are validated and more fully understood. In the Philippines, the IPM CRSP team works closely with local farmers in San José as well as the more prosperous onion farmers of Bongabon, Nueva Ecija, most of whom are members of NOGROCOMA. Bangladesh farmer-scientist cooperative projects take place in intensive vegetable-growing areas in the Gazipur, Jessore, Rangpur, and Comilla districts. Taking stakeholder input into account, IPM CRSP scientists propose specific experiments with clear objectives, testable hypotheses, descriptions of activities, and justification and relationship of the proposed activities to other IPM CRSP activities at the site. Expected outputs and impacts are also described. If the activity is continuing from the previous year, a brief project update is included. Proposed experiments are presented to the IPM CRSP team, including in-country and cooperating institution scientists, early in the calendar year. In this way all proposals are reviewed by an interdisciplinary group representing a broad range of expertise and interests, insuring that proposals are critically and thoroughly reviewed. All proposals are reviewed by the full Technical Committee of the IPM CRSP prior to final approval of the workplan. The formal process of proposing and designing experiments and compiling them into a workplan is carried out once each year.

Research, extension, and training activities in both sites focus on developing environmentally and economically sound approaches to managing pests in rice-vegetable and vegetable-vegetable cropping systems, with focus on onion and eggplant in the Philippines and eggplant (brinjal), tomato, cucurbits, okra, and cabbage in Bangladesh. Research studies to determine both immediate and long-range solutions to the most critical pest problems are conducted through: 1) multidisciplinary on-farm studies; 2) multidisciplinary laboratory, greenhouse, and microplot studies; 3) socio-economic impact analysis; and 4) IPM technology transfer. The activities, integrated among the various disciplines (entomology, plant pathology, weed science, nematology, sociology, economics) address a broad range of IPM strategies from validating indigenous farmers' cultural practices to chemical, biological, or genetic methods. While most studies employ a single-crop approach (vegetables only), multi-crop approaches are used on pests such as purple nutsedge that appear in both rice and vegetable crops. To encourage the active participation of farmers in the IPM approach and to enhance interaction between researchers and farmers, most field studies are conducted in farmer-cooperators' fields. Socio-economic impact analyses are an integral part of the program, and relevant economic data are collected for each study.

From results of completed studies, promising technologies, including pheromone trapping for cutworms in onions and weed management through optimized hand weeding and reduced herbicide use, are being integrated into the IPM training programs of the PhilRice Extension and Training Division in the Philippines. The training programs are targeted to training of trainers (provincial level) who in turn train agricultural technicians (municipal level). Mature technologies are also shown in techno-demo plots in village-level integration studies conducted in pilot areas. In Bangladesh, several effective IPM technologies, such as mass trapping of fruit flies in cucurbits, the use of poultry compost-based soil amendments to reduce soil-borne diseases, and grafting of eggplant and tomato onto bacterial wilt-resistant rootstocks have been adopted by farmers in the districts in which initial field trials were established, primarily by "word of mouth" and through the efforts of farmer-leaders/early adopters working closely with IPM CRSP scientists. IPM CRSP scientists and technical staff cooperated with CARE to train primarily female farmers with small land-holdings in several additional districts in the grafting technology.

Integrating Social, Gender, and Economic Analysis

Gender and social impact analysis have been important components of the IPM CRSP program in the Asia sites, especially in the Philippines. PAs and baseline surveys helped identify gender roles in production and decision-making within the households. For example, in the Philippines, it was found that land preparation and threshing are male activities while transplanting, weeding, sorting seedlings, harvesting, and grading onions are primarily completed by women. Women manage the family budget, obtain credit from informal sources for both the household and farm, make marketing decisions, seldom spray pesticides but are interested in practices that might reduce expenditures on pesticides, and can identify symptoms of certain pests. Generally, power-intensive operations were handled by men and control-intensive ones by women. The importance of involving both women and men in IPM was evident and was taken into account when designing farmer training. In Bangladesh, women had a smaller role in decision-making in the household in general, but were involved in pest management, including decisions to apply pesticides.

Gender and social impact analysis (SIA) have been used to assess the likely direct and indirect benefits of IPM technologies and to understand factors affecting adoption and diffusion. For example, social impact assessment of the practice of rice hull burning on fields prior to onion production in the Philippines indicated that the technology is widely adopted in regions where rice hulls are readily available, due to farmers' perceptions that the practice reduces weeds, improves soil fertility, and increases the number of Yellow Granex onions qualifying as export grade. Negative impacts or constraints were primarily related to environmental and health effects from smoke and increased competition for rice hulls (Roguel et al., 2003).

Researchers evaluated farmers' perceptions of sex pheromone traps as a monitoring device to determine the time of insecticide application for *Spodoptera litura* and *Spodoptera exigua* on onions. Farmers were enthusiastic about trying the technology in their own fields, and they perceived no adverse effects on their health and the environment. With proper training and an information campaign, the use of sex pheromone traps can significantly reduce the use of insecticides. However, farmers perceive the traps will not reduce their insecticide application if other pest problems are not addressed. They believe that the other insects still present in their fields will require insecticides for "preventive" measures, implying that the component technology may be less successful absent a more complete IPM program. They also stress the importance of the technology being readily accessible and adopted on a community-wide level.

In addition to the social impact assessments, economic impact assessments were conducted for virtually every component of the IPM program at the two sites. Partial budget forms were prepared in each country for use by scientists as they conduct experiments in the fields, as budgeting is the first step in assessing the economic viability of alternative IPM practices. In some cases these IPM technologies were used to project aggregate economic benefits, both direct and indirect, due to the value of health and environmental benefits (See Chapters 12 and 13 in this volume). Policy analysis was also conducted with respect to certain issues such as net subsidies or taxes on pesticides that might be providing incentives or disincentives to adopt IPM (See Chapter 10 in this volume).

Role of and Means for Facilitating Interdisciplinary Approaches

Successful IPM programs require an interdisciplinary approach, taking into account not only the management of weeds, diseases, and insects, but also the economic and social impacts of such programs. The need for this approach stems from both biological interactions and social and economic factors that influence IPM decision-making. Interdisciplinary approaches are facilitated in both Asia sites by the formation of teams of scientists from in-country, the United States, and other cooperating institutions comprised of individuals with strong disciplinary focus but also a willingness to plan and participate in experiments designed according to a systems paradigm. The planning process described previously promotes the systems approach and maximizes interactions among scientists. Experiments are conducted in an interdisciplinary mode where feasible. For example, in the Philippines, the effects of rice-hull burning on weed populations, root knot nematode, and pink root disease in onion were evaluated in a series of experiments (Baltazar et al., 2000; Gergon et al., 2001b). In other cases, component research may be done, and the most suitable lessons learned combined in later studies, from testing "best-management packages" to on-farm research and demonstration experiments, such as the Village Level Integration studies conducted in the Philippines (Baltazar et al., 2002). In Bangladesh, multidisciplinary teams have been formed to carry out evaluation of best pest-management packages for tomato, cucurbit crops, and cabbage.

In both the Asian sites, one key to successful interdisciplinary interaction has been to focus all scientists not just on the science but on the goals of increased profitability and maximizing IPM adoption to generate widespread gains. A second has been to resolve personality conflicts wherever possible. Interdisciplinary work requires respecting other disciplines and learning enough about the needs of the other disciplines to communicate well. A third key has been to support and utilize a strong coordinator who understands scientific field work, institutional constraints, and the need to integrate social scientists.

Facilitating Interactive Transnational Linkages, Including Roles of NARS, IARCs, and Universities

In both IPM CRSP sites in Asia, the principal home-country cooperating institution is a national agricultural research center. For the Philippines site, PhilRice serves as the principal contractor with Virginia Tech under a Memorandum of Agreement between the two institutions. This structure allows the project access to well-trained PhilRice scientists, as well as some of the best research facilities available. From early in the project, formal relationships were also forged with the University of the Philippines at Los Baños, the premier academic program for agriculture in the Philippines. Two of the four scientists who have served as site coordinators for the Philippines are faculty members of UPLB on leave from their home departments while coordinating the IPM CRSP through PhilRice. Other participating universities in the Philippines are Central Luzon State University (CLSU) and Leyte State University. Faculty from these institutions collaborated on the project on study leave at PhilRice and later as cooperators, and through the training of graduate students. Both the AVRDC and IRRI have played a critical role in developing and conducting research programs through direct participation of scientists, introduction of technologies and germplasm, and short-term training of Philippine scientists. In Bangladesh, activities are directed from the Horticultural Research Center (HRC) of BARI in Gazipur, under the auspices of the Bangladesh Agricultural Research Council. IPM CRSP team members from BARI cooperate with faculty from Bongabandu Sheik Mujibur Rahman Agricultural University (BSMRAU) and UPLB, and scientists from IRRI and AVRDC. Participation from the United States is multi-institutional, with the active involvement of faculty from Pennsylvania State University, Ohio State University, Virginia Tech, Purdue University, and the University of California–Davis. Linkages between the South and Southeast Asia sites are reinforced by the involvement of scientists from U.S. universities, UPLB, and AVRDC in both sites. Students from the Philippines and Bangladesh have been trained in M.S. and Ph.D. graduate degree programs in cooperating U.S. institutions and UPLB in economics, sociology, statistics, weed science, and entomology. Other students participated in sandwich programs in which course work or research training was carried out in the United States, while degrees were granted from home institutions including UPLB and BSMRAU. Finally, many short-term training opportunities were provided for researchers and students in one of the U.S. cooperating institutions, UPLB, or AVRDC.

Examples from Bangladesh and the Philippines – Developing IPM Systems

Management of Eggplant Fruit and Shoot Borer

Eggplant fruit and shoot borer (EFSB), *Leucinodes orbonalis* (see Figure 3-1, page xviii) is perhaps the single most important pest of eggplant in Asia. Farmers may apply insecticides 50 times or more in a cropping season, often with little success. In eggplant-intensive growing regions in Bangladesh, 60% of farmers apply insecticides more than 141 times in a single growing season (Rashid et al., 2003). In spite of pesticide applications, more than one-third of eggplant production in Bangladesh is lost due to EFSB damage (BARI, 1999). The practice of frequent pesticide applications over the years has not only complicated EFSB management by disrupting natural control systems, but has also created health problems to applicators and has increased cultivation costs substantially. On average, about 29% of the total production cost for eggplants in Bangladesh is due to pesticide purchases (Rashid et al., 2003).

Data from on-farm studies in the Philippines indicate that simply removing damaged fruits and shoots at weekly intervals can reduce EFSB infestations (Table 3-1). When done at the same time as harvest, this practice reduced labor costs and resulted in a net incremental benefit of

Table 3-1. Effects of cultural and chemical methods on control of fruit and shoot borer and eggplant yield, San Jose, Nueva Ecija, the Philippines, 1997 dry season[1].

TREATMENT	Frequency (no/wk)	Total fruits (no/20m^2)	Undamaged fruits (kg/20m^2)	Total no. of larvae (no/20m^2)
Chlorpyrifos	2/wk	584 a	33 c	110 a
Chlorpyrifos	1/wk	626 a	34 c	94 ab
Chlorpyrifos	1/2wks	645 a	34 c	101 ab
Chlorpyrifos	1/3wks	671 a	42 ab	81 b
Fruit/shoot removal		687 a	43 a	32 c
Untreated control		662 a	37 bc	120 a

[1]Means in a column followed by a common letter are not significantly different at $P \leq 0.05$.

$2500/ha for weekly removal and $1000/ha for biweekly removal. Spraying an appropriate insecticide every 3 weeks also reduced fruit and shoot borer infestation comparable to weekly removal of damaged fruits (Table 3-1). Adoption of a significantly reduced insecticide application approach would reduce insecticide use from 30-50 to only six in a cropping season, implying tremendous reduction in production costs as well as reduced risks to human health and the environment.

A second approach in EFSB management is the development or identification of pest-resistant eggplant varieties. Field evaluations in Bangladesh of local and exotic germplasm, carried out for three years at the BARI farm, led to the identification of twelve cultivars having high to moderate EFSB resistance. Two of the selected cultivars are recommended varieties (Kazla and Uttara), both of which have moderate resistance (Table 3-2).

Selected resistant materials were used as donors in developing eggplant varieties having improved agronomic qualities and pest-resistance traits. Two resistant lines, BL-009 and BL-114, which have attractive agronomic characters, are now being demonstrated in farmer fields to assess their acceptability. As an immediate measure, farmers are being advised to grow the moderately EFSB-resistant varieties Kazla and Uttara.

In field surveys, indigenous populations of EFSB parasitoids were low, mainly due to frequent pesticide application. However, the larval and pupal parasitoid *Trathala flavoorbitalis* was well distributed in eggplant-production areas. Studies in two distant districts of Jessore and Gazipur showed that

Table 3-2. Field resistance of eggplant varieties/lines to eggplant fruit and shoot borer.

	Shoot infestation (%)[1]		Fruit infestation (%)[2]	
Variety/line	2000-2001	2001-2002	2000-2001	2001-2002
BL-107	0.4 (HR)	8.3 (R)	0.4 (HR)	0.2 (HR)
EG-195	0.8 (HR)	10.3 (R)	0.7 (HR)	0.7 (HR)
TS-060B	0.0 (HR)	6.7 (R)	0.0 (HR)	0.9 (HR)
BL-072	—	3.3 (HR)	13.4 (MR)	0.9 (HR)
BL-095(2)	—	7.8 (R)	12.6 (MR)	0.7 (HR)
BL-009	—	4.7 (HR)	8.4 (R)	0.9 (HR)
BL-095	1.1 (HR)	8.3 (R)	—	1.2 (R)
BL-114	—	4.7 (R)	14.5 (MR)	1.6 (R)
EG-203	—	7.3 (R)	0.5 (HR)	3.0 (R)
Kazla	1.9 (R)	7.3 (R)	3.9 (R)	7.1 (MR)
Uttara	—	14.3 (R)	—	14.3 (MR)

[1] Reactions based on infestation rates of <5% = Highly Resistant (HR); <15% = Resistant (RO; <30% = Moderately Resistant (MR); <50% = Susceptible (S); >50% = Highly Susceptible (HS).

[2] Based on infestation rates of <1% = HR; <10% = R; <20% = MR; <40% = S; >40% = HS.

populations of this parasitoid increased about ten-fold within a year when no insecticides were applied. Parasitism of EFSB by *T. flavoorbitalis* increased about three-fold after one year of eggplant cultivation without insecticide use. The incidence pattern of the parasitoid was dependent on that of its host; a significant correlation between the number of EFSB and the parasitoid was observed at both Gazipur ($r = 0.87$, $p < 0.05$) and Jessore ($r = 0.71$, $p < 0.05$). In greenhouse tests, EFSB infestation of 1st instar larvae decreased by 91% in the presence of the parasitoid. Similar results were obtained in microplot tests with one susceptible and 11 resistant varieties. Except for three varieties, EFSB infestations in the presence of the parasitoid decreased by 45 to 66% (Table 3-3).

Management of Bacterial Wilt in Eggplant

Bacterial wilt (BW) disease is caused by *Ralstonia solanacearum*, a soil-borne bacterium that can rapidly disperse within and among fields through cultivation processes. Strains in the Philippines that attack eggplant have been identified as race 1 biovars 3 and 4; survey data on the strains present in Bangladesh are not available, but race 1 biovar 3 has been identified (Opina et al., 2001). Farmers lose more than 50% of eggplant production

Table 3-3. Eggplant fruit and shoot borer (EFSB, *Leucinodes orbonalis*) infestation rates on resistant and susceptible varieties/lines in the presence or absence of parasitoids.

	Shoot infestation (%) by EFSB	
Variety/line[1]	With parasite[2]	Without parasite
BL-107 (HR)	2.8 (63%)	8.5
EG-195 (HR)	4.9 (55%)	11.0
TS-060B (HR)	6.3 (3%)	6.5
BL-072 (HR)	4.5 (0%)	4.2
BL-095(2) (HR)	3.5 (59%)	8.6
BL-009 (HR)	4.8 (8%)	5.2
BL-095 (R)	3.9 (61%)	10.1
BL-114 (R)	3.5 (45%)	6.4
EG-203 (R)	5.7 (46%)	8.9
Kazla (MR)	3.1 (63%)	8.3
Uttara (MR)	5.9 (66%)	17.4
EG-075 (S)	10.3 (59%)	25.4

[1] Letters in parentheses indicate reactions to EFSB: HR = Highly resistant; R = Resistant; MR = Moderately resistant; and S = Susceptible.
[2] Percent reduction of infestation as a result of parasitism.

in Bangladesh due to BW attack in areas where the disease is endemic. Yield losses in the Philippines consistently reach 30 to 80% in Central Luzon. Grafting BW-susceptible eggplant scions onto BW-resistant rootstocks has been demonstrated to be effective in managing bacterial wilt in both countries. Research in Bangladesh demonstrated that the wild *Solanum* species, *S. torvum* and *S. sisymbriifolium*, are highly resistant to the disease in intensive vegetable-production areas. A simple, affordable polyhouse (Fig. 3-2) was designed in which seedlings were grafted successfully (>90% survival rate) using the cleft/clip technique. In a 2002 field trial in 13 fields in two villages in Jessore, an intensive vegetable-production area with a history of losses due to bacterial wilt, eggplant 'Chega' grafted onto *S. torvum* and *S. sisymbriifolium* had significantly lower mortality, longer harvest duration, and higher yield than non-grafted 'Chega' (Tables 3-4, 3-5). Subsequently, net income to farmers from these fields averaged 2.8 – 4 times higher than in fields with non-grafted plants (Table 3-5). Mortality among grafted plants was almost exclusively due to Phomopsis stem rot, while the majority of loss among the non-grafted plants was due to bacterial wilt.

Figure 3-2. Polyhouse (Jessore, Bangladesh) designed and constructed for local production of grafted eggplant and tomato for management of bacterial wilt caused by Ralstonia solanacearum.

In the Philippines, a bacterial wilt-resistant *Solanum melongena* line, EG203, identified by AVRDC, was utilized in similar experiments. Results of on-farm studies in Central Luzon showed that grafting of susceptible commercial cultivars to EG-203, a resistant cultivar developed at AVRDC, increased resistance to bacterial wilt by 20 to 30%. These cultivars yielded higher when grafted to EG-203, compared to non-grafted plants.

Table 3-4. Plant mortality and harvest duration for grafted and non-grafted eggplant (brinjal) in fields infested with *Ralstonia solanacearum* in two villages in Jessore, Bangladesh.

Location (No. of fields)	Rootstock/Scion	Mortality (%)	Harvest Duration (days)
Naodagagram, Jessore (7)	*S. torvum*/Chega	10.5 b	90 a
	S.sisymbriifolium/Chega	8.1 c	89 a
	Chega, non-grafted	31.6 a	60 b
Gaidghat, Jessore (6)	*S. torvum*/Chega	10.5 b	84 a
	S. sisymbriifolium	7.5 c	77 b
	Chega, non-grafted	27.3 a	55 c
Sripur, Gazipur (1)	*S.torvum*/Singnath	7.2	86
	S.sisymbriifolium/Singnath	8.8	80
	Singnath, non-grafted	100.0	—

Table 3-5. Yield and net income for grafted and non-grafted eggplant (brinjal) in fields infested with *Ralstonia solanacearum* in two villages in Jessore, Bangladesh.

Location (No. of fields)	Rootstock/Scion	Yield (t/ha)	Increase (%)	Net Income ($ US/ha)	Increase (%)
Naodagagram, Jessore (7)	*S. torvum*/Chega	32 a	246	2177	286
	S. sisymbriifolium/Chega	33 a	254	2271	299
	Chega	13 b	—	760	—
Gaidghat, Jessore (6)	*S. torvum*/Chega	27 b	245	1668	327
	S. sisymbriifolium	31 a	282	2032	398
	Chega	11 c	—	510	—
Sripur, Gazipur (1)	*S. torvum*/Singnath	27	—	1639	—
	S. sisymbriifolium/Singnath	30	—	1948	—
	Singnath	—	—	—	—

Integrated Weed Management in Onion in the Philippines.

The sedge *Cyperus rotundus* (purple nutsedge) and the broadleaf weed *Trianthema portulacastrum* (horse purslane) are the most important competitors of onion in the Philippines (Baltazar et al., 1999a) (see Figure 3-3, page xviii). The farmers' practice of applying one to two herbicides followed by hand-weeding (Figure 3-4) one to three times for adequate season-long control can cost as much as $300/ha, 20% of total production costs (Baltazar et al., 1999b). IPM CRSP on-farm studies consistently showed that one application of the correct herbicide followed by one timely hand weeding controlled weeds, with onion yields comparable to yields in onions managed according to the farmers' practice (Baltazar et al., 2000). Weed-control costs were reduced by 15 to 70% without reducing weed-control efficacy or yields, resulting in an average net incremental benefit of $500/ha compared to the farmers' practice. Another weed-management technique, rice-straw mulching, was validated in on-farm studies as a method to reduce weed growth by 60%, increase yields by 70%, and reduce weed-control costs by more than 50% over non-mulched plots (Baltazar et al., 2000). This technique, an indigenous cultural practice in multiplier onion (shallot), may also be applicable to other onion types.

Perennial weeds, including purple nutsedge, have become increasingly adapted to rice-onion production systems, and tuber populations have increased over time due to the continuous rice-onion rotation cycle

Figure 3-4. Hand-weeding onions in the Philippines.

(Casimero, 2000; Casimero et al., 1999, 2001). As a consequence, multi-season management approaches are most cost-effective in reducing weed populations. One such approach, the stale-seedbed technique (sequential harrowing or harrowing followed by a non-selective herbicide at biweekly intervals), done during fallow periods between the rice and onion crop, reduced purple nutsedge tuber populations by 80 to 90% over four cropping seasons in 2-year on-farm studies (Figure 3-5) (Baltazar et al., 2001). This decreased hand-weeding costs by 40 to 50%, increased yields by 2 t/ha, and increased net incomes by $1000 over those of farmers' practice (Table 3-6).

Conclusions

The IPM CRSP research and development efforts described in this chapter have contributed to filling knowledge gaps in IPM for vegetable crops in Asia. Some of the most difficult and economically damaging pest problems were addressed and technologies developed and disseminated to farmers, where they are in place to reduce losses due to diseases, insect pests, and weeds. At the same time, utilization of these technologies also reduces potential harmful effects to farmers, consumers, and the environment from the overuse and/or misuse of pesticides. Through an interdisciplinary, participatory approach to IPM research involving multiple stakeholders, pest-management issues were identified and addressed in a way that would maximize the probability of adoption by farmers.

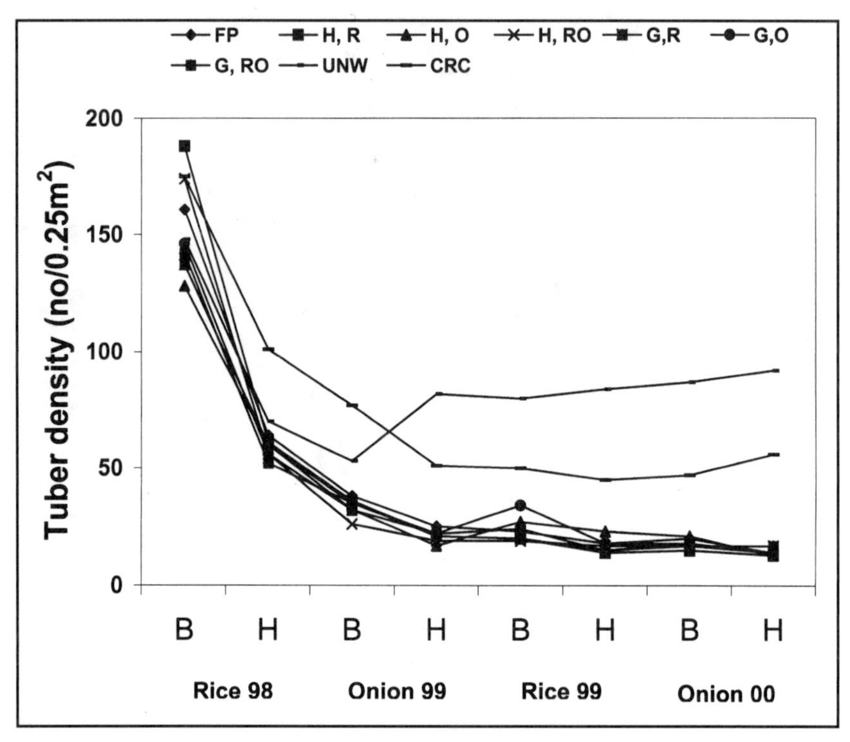

*Figure 3-5. Reduction in purple nutsedge (*Cyperus rotundus*) tuber populations resulting from multi-season interventions. FP=Farmer's Practice; H,R=Harrow, before rice; H,O=Harrow, before onion; H,RO=Harrow, before rice and onion; G,R=Glyphosate before rice; G,O=Glyphosate before rice and onion; UNW=Unweeded control; CRC=* Cyperus rotundus *control.*

Economic and social impact analyses were carried out to insure potential profitability and social acceptance of new technologies. In some cases, technologies that were shown to mitigate pest problems were rejected because of a low probability of adoption due to economic or social factors. For example, experiments in the Philippines showed that construction of a net barrier around vegetable crops maintained low pest pressure and eliminated the need for pesticide application, but costs of construction and maintenance of the structure were prohibitive. In another instance, rice-hull burning, an indigenous practice in parts of the Philippines, was effective in reducing weeds, root knot nematode, and pink root disease in onion, but social impact assessment indicated that it would be unlikely to be widely adopted due to air pollution concerns and inadequate access to rice hulls in

Table 3-6. Net income[1] in rice and onion crops treated with stale seedbeds from 1998 to 2000 wet and dry seasons, San Jose, Nueva Ecija, the Philippines.

	Net Income ($/ha)[2]				
Treatment	Rice 98	Onion 99	Rice 99	Onion 00	Total
Farmer's practice	539	936	3154 (83)	2664 (65)	7293
Harrow (rice)	501	1002	3220 (84)	2544 (61)	7268
Harrow (onion)	392	946	3718 (89)	3086 (69)	8141
Harrow (both)	458	960	3804 (88)	3026 (68)	8248
Glyphosate (rice)	524	964	3317 (84)	2664 (64)	7470
Glyphosate (onion)	331	926	3578 (91)	3045 (69)	7880
Glyphosate (both)	615	1025	3618 (83)	3091 (67)	8350
Unweeded	767	841	2375 (68)	1760 (56)	5743
C. rotundus alone	532	926	3039 (87)	2097 (97)	6594

[1] Gross income minus weed control costs (partial budgeting, does not include other production costs). Farmgate price of rice: $0.25/kg (1998); $0.21/kg (1999). Farmgate price of onion cv. Tanduyong: $0.25/kg (1998); $0.20/kg (2000).
[2] Figures in parentheses indicate percent increase over first crop.

onion-intensive areas. Successful technologies such as disease- and insect-resistant varieties, grafted tomato and eggplant for bacterial wilt resistance, pheromones and/or bait traps to manage insect pests in cucurbits and onion, and various weed-management strategies, including the stale seedbed technique and reduced herbicide use plus hand weeding, are being adopted by farmers through training efforts carried out in both countries and by word-of-mouth. IPM CRSP scientists in both Bangladesh and the Philippines have forged strong ties with local, regional, and national outreach organizations, as well as NGOs and farmer-based organizations, to disseminate promising technologies as widely as possible.

Acknowledgments

The contributions of the following scientists and technical staff to this project are gratefully acknowledged: Philippines: R.T. Alberto, G.S. Arida, M. Brown, M.A.A. Capricho, M.T. Caasi-Lit, M.C. Casimero, B. Catudan, S.K. De Datta, Q.D. Dela Cruz, M.V. Duca, D.T. Eligio, S.R. Francisco, R.M. Gapasin, V.P. Gapud, E.B. Gergon, R.G. Maghirang, R.B. Malasa, E.C. Martin, S.S. Mina-Roguel, S.R. Obien, N.L. Opina, L.E. Padua, N. Perez, B.S. Punzall, J.M. Ramos, C.C. Ravina, J. Recta, S.E. Santiago, L.S. Sebastian, I.R. Tanzo, and E.R. Tiongco; Bangladesh : M. Akand, M.S.

Alam, S.N. Alam, A. Bashet, M.I. Faruque, A. Hossain, M.M. Hossain, S. Hossain, Z. Islam, H.S. Jasmine, K.A. Kader, A. Karim, S.A. Khan, M. Khorseduzzaman, M.A.T. Masud, S.R. Mollik, M. Naziruddin, A.K.M. Quamaruzzaman, M. Rafiquddin, N.S. Nahar, A. Rahman, M.A. Rahman, M. Rahman, M.A. Rashid, M.H. Rashid, L. Yasmin, M.F. Zahman; AVRDC: L.L. Black, N.S. Talekar; IRRI: T. Mew, K.L. Heong, A.M. Mortimer; and U.S. Institutions: R. Gilbertson (Univ.California-Davis), S. Hamilton (Virginia Tech), C. Harris (Virginia Tech), L.T. Kok (Virginia Tech), G. Luther (Virginia Tech), C. Sachs (Penn State), G. Shively (Purdue). The Southeast Asia site also thanks Mrs. Dulce Gozon and the NOGROCOMA onion growers' cooperative for outstanding and continuous support of this program.

References

Baltazar, A.M., R.T. Alberto, L.E. Padua, M.C. Lit, N.L. Opina, G.S. Arida, E.B. Gergon, M.C. Casimero, S.R. Francisco, I.R. Tanzo, S.A. Miller, G.W. Norton, E.G. Rajotte, S.K. De Datta, L.S. Sebastian, V.P. Gapud, and S.R. Obien. 2002. IPM in rice-vegetables systems: Research highlights and promising technologies. *PhilRice Technical Bulletin* 6(2): 66-71.

Baltazar, A.M., E.C. Martin, M.C. Casimero, F.V. Bariuan, S.R. Obien, and S.K. De Datta. 1999a. Major weeds and dominance patterns in rainfed rice-onion cropping systems. *The Philippine Agricultural Scientist* 82(2): 167-178.

Baltazar, A.M., E.C. Martin, M.C. Casimero, F.V. Bariuan, S.R. Obien, and S.K. De Datta. 1999b. Weed management strategies in onion grown after rice. *Proceedings, 17th Asian-Pacific Weed Science Society Conference*, November 22-27, 1999, Bangkok, Thailand.

Baltazar, A.M., E.C. Martin, M.C. Casimero, F.V. Bariuan, S.R. Obien, and S.K. De Datta. 2000. Reducing herbicide use with agronomic practices in onion (*Allium cepa*) grown after rice (*Oryza sativa*). *The Philippine Agricultural Scientist* 83(1): 34-44.

Baltazar, A.M., E.C. Martin, A.M. Mortimer, M.C. Casimero, S.R. Obien, and S.K. De Datta. 2001. Reducing *Cyperus rotundus* populations using stale-seedbed techniques in rice-onion cropping systems. *Proceedings 18th Asian-Pacific Weed Science Society* conference, May 28-June 2, 2001, Beijing, China.

Bangladesh Agricultural Research Institute (BARI). 1999. *Annual Report 1998–1999*. Joydebpur, Gazipur, Bangladesh: BARI. 482 pp.

Campinera, J. 2001. *Handbook of Vegetable Pests*. Academic Press. 460 pp.

Casimero, M.C. 2000. Population dynamics, growth and control of weeds in rainfed rice-onion cropping systems. Ph.D. Thesis. University of the Philippines Los Banos College, Laguna. 139 pp.

Casimero, M.C., A.M. Baltazar, J.S. Manuel, S.R. Obien, and S.K. De Datta. 1999. Morphologic and genetic variations in upland and lowland ecotypes of purple nutsedge in rainfed rice-onion systems. *Proceedings 17th Asian-Pacific Weed Science Society* Conference, November 22-27, 1999, Bangkok, Thailand. 17: 134-139.

Casimero, M.C., A.M. Baltazar, J.S. Manuel, S.R. Obien, and S.K. De Datta. 2001. Population dynamics and growth of weeds in rice-onion systems in response to chemical and cultural control methods. *Proceedings 18th Asian-Pacific Weed Science Society* conference, May 28-June 2, 2001, Beijing, China.

Gapasin, R., M.V. Judal, C. Pile, E.B. Gergon, V. Gapud, and S.R. Obien. 1998. Alternative management strategies against rice root-knot nematode, *Meloidogyne graminicola,* in rice-onion system. *Philippine Technology Journal* 23: 69-75.

Gapud, V.P., R.T. Alberto, S.K. De Datta, S.R. Obien, A.M. Baltazar, L.E. Padua, M.C. Lit, N.L. Opina, G.S. Arida, E.B. Gergon, M.C. Casimero, G.W. Norton, S.A. Miller, and E.G. Rajotte. 2000. IPM in rice-vegetable systems: the IPM CRSP experience. *Proceedings, 31st Pest Management Council of the Philippines Conference*, Baguio City, Philippines, 3-6 May, 2000.

Gergon, E.B., S.A. Miller, and R.G. Davide. 2001a. Root-knot disease of onion caused by the rice root-knot nematode *Meloidogyne graminicola*: Occurrence, pathogenicity, and varietal resistance. *Philippine Agricultural Scientist* 84:43-50.

Gergon, E.B., S.A. Miller, R.G. Davide, O.S. Opina, and S.R. Obien. 2001b. Evaluation of cultural practices (surface burning, deep ploughing, organic amendments) for management of rice root-knot nematode in rice-onion cropping system and their effect on onion. *International Journal of Pest Management* 47: 265-272.

Gergon, E.B, S.A. Miller, and R.G. Davide. 2001. Effect of rice root-knot nematode on growth and yield of onion. *Plant Disease* 86: 1339-1344.

Guan-Soon, L. 1990. Overview of vegetable IPM in Asia. *FAO Plant Protection Bulletin* 38: 73-88.

Hossain, M.I., S.M.N. Alam, M.A.M. Miah, and M.N. Islam. 2002. Diffusion of IPM technologies for vegetables cultivation: Empirical results from farmers' experiences. Paper presented at the Asia Regional Conference on Public-Private Sector Partnership for Promoting Rural Development, Bangladesh Agricultural Economists Association and International Association of Agricultural Economists, October 2-4, 2002, Dhaka, Bangladesh. 10pp.

Hossain, M.I., G.E. Shively, and G.W. Norton. 1999. Baseline survey of existing patterns of vegetable production and pest management practices in farmers' fields. *IPM CRSP Sixth Annual Report, 1998-1999,* IPM CRSP, Office of International Research and Development, Virginia Tech, Blacksburg, Va., pp. 403-407.

IPM CRSP. 1999. IPM CRSP Update (Newsletter), Volume 5, no. 1. Office of International Research and Development, Virginia Tech, Blacksburg, Va. 4pp.

Islam, Z., G.J. Uddin, M.A. Nahar, A.H. Rahman, M.S. Nahar, M.A. Rahman, M.S. Hasan, L. Black, E. Rajotte, A. Baltazar, S.K. De Datta, and G. Luther.

1999. Monitoring of crop pests and their natural enemies in rice and vegetables in rice-vegetable cropping systems. *IPM CRSP Sixth Annual Report, 1998-1999,* IPM CRSP, Office of International Research and Development, Virginia Tech, Blacksburg, Va., pp. 407-412.

Litsinger, J., G.W. Norton, and V. Gapud. 1994. Participatory appraisal for IPM research planning in the Philippines. IPM CRSP, Office of International Research and Development, Virginia Tech, Blacksburg, Va. 130 pp.

Lucas, M.P., S. Pandey, R.A. Villano, D.R. Culanay, and S.R. Obien. 1999. Characterization and economic analysis of intensive cropping systems in rainfed lowlands of Ilocos Norte, Philippines. *Experimental Agriculture,* 35, no. 2 (April 1999): 211-224.

Ooi, P.A.C., and J.K. Waage. 1994. Biological control in rice: applications and research needs. Pp. 209-216 in *Rice pest science and management: selected papers from the International Rice Research Conference,* ed. P.S. Teng, K.L. Heong, K. Moody. International Rice Research Inst., Los Banos, Laguna (Philippines). Los Banos, Laguna (Philippines). IRRI. 1994.

Opina, N.L., R.T. Alberto, S.E. Santiago, L.L. Black, and S.A. Miller. 2001. Influence of host resistance and grafting on the incidence of bacterial wilt of eggplant. *IPM CRSP Eighth Annual Report 2000-2001.* Virginia Tech, Blacksburg, Va., pp. 21-23.

Rahman, S. 2003. Farm-level pesticide use in Bangladesh: Determinants and awareness. *Agriculture Ecosystems and Environment* 95: 241-252.

Rashid, M.A., S.N. Alam, F.M.A. Rouf, and N.S. Talekar. 2003. Socio-economic parameters of eggplant pest control in Jessore District of Bangladesh. Shanhua, Taiwan: AVRDC – The World Vegetable Center, AVRDC Publication No. 03-556. 29 pp.

Roguel, S.M., R.B. Malasa, and I.R. Tanzo. 2003. Social impact assessment of weed management strategies in rice-onion systems. *Proceedings 19th Asian-Pacific Weed Science Society Conference.* March 17-21, 2003, Manila, Philippines, pp. 883-890.

Savary, S., F.A. Elazegui, K. Moody, J.A. Litsinger, and P.S. Teng. 1994. Characterization of rice dropping practices and multiple pest systems in the Philippines. *Agricultural Systems* 46: 385-408.

— 4 —
Developing IPM Packages in Africa

J. Mark Erbaugh, John Caldwell, Samuel Kyamanywa,
Kadiatou Toure Gamby, and Keith Moore

Farmer participation and integrated pest management (IPM) are growing and complementary components of agricultural research and extension programs in sub-Saharan Africa. Over several decades, attempts to develop and disseminate IPM throughout the continent met with limited success (Yudelman et al., 1998; Morse and Buhler, 1997; Kiss and Meerman, 1991). However, increasing farmer participation in the development and implementation of IPM programs has emerged as a promising strategy for increasing the relevance and use of IPM, particularly among small-scale farmers (Yudelman et al., 1998; Dent, 1995).

Multiple approaches have guided IPM program implementation since its inception, resulting in different research and extension strategies (Ehler and Bottrell, 2000; Yudelman et al., 1998). One, the *tactical,* economic-threshold approach to IPM, evolved in response to environmental concerns about the overuse of synthetic pesticides as the sole control tactic in intensive-input agricultural systems in developed countries. Its goal was to develop a mix of pest and disease-control tactics in which synthetic pesticides were used as one control tactic defined by economic thresholds of pest populations. IPM programs in the United States used this approach for more than 30 years, and, for many farmers, IPM is equated with thresholds. The role of extension in this approach was to directly transfer and disseminate these tactics and the use of thresholds to farmers (Morse and Buhler, 1997).

Alternative approaches have been developed in recent years, including alternatives for small-scale farming systems in sub-Saharan Africa (SSA), which are often characterized by the absence or minimal use of chemical pesticides, as well as of other synthetic production inputs. These newer approaches in SSA seek to combine indigenous farmer knowledge with scientific knowledge of agro-ecologies and pests to develop site-specific IPM systems. Variously labeled as ecological, strategic, or sustainable IPM (Pimbert, 1991; Mangan and Mangan, 1998; Schwab et al., 1995), these approaches are often described as being knowledge-intensive (Morse and Buhler, 1997) because they often require enhanced knowledge and understanding of biological factors and site-specific ecological interactions for their successful implementation by small farmers (Dent, 1995). Since information and knowledge exchange between farmers and scientists is an essential objective, ecological IPM programs are increasingly linked to participatory research and extension approaches (Norton et al., 1999).

Participatory approaches have gained ascendance not only with IPM programs but also with a wide variety of development efforts in sub-Saharan Africa, including public health, community development, and microfinance programs. In sub-Saharan Africa, one of the most important obstacles to agricultural development is a lack of appropriate technological solutions for diverse agricultural production systems (Cleaver, 1993). Participatory strategies are premised on the recognition that knowledge possessed by local inhabitants is not only legitimate in terms of traditional systems, but can also contribute to the development of new systems, enhancing the relevance and applicability of knowledge from formal scientific research. Since farmers co-evolved with their present agro-ecosystems, their cumulative knowledge and experience with crop production represents a source of valuable information and a foundation on which to construct and advance new solutions and technologies. Thus, farmer participation in agricultural research activities in sub-Saharan Africa represents an effective mechanism for developing and adapting farmer and scientist co-generation of knowledge to diverse conditions.

The need for strategic approaches also reflects a fundamental developmental characteristic of African agricultural systems. In contrast to input-intensive systems, which are fully integrated into the market economy, many African agricultural systems are in transition. Changing levels of agricultural development from formerly extensive and semi-subsistence-oriented production systems to more intensive and partially commercialized production systems characterize this transition. In other parts of the world, this transition has led to changing agricultural practices including input-intensification, induced a changing set of pest problems, and resulted in increased use of synthetic pesticides. A strategic approach to IPM seeks to develop an alternative trajectory for this transition that does not lead to the same outcome of dependence on pesticides. The strategic approach also recognizes that different regions and crops are at different stages on the transitional continuum, as illustrated with concrete examples in Box 1. The implications for an IPM research program are that the socio-economic and production context, as defined by the level of agricultural development, will affect pest priorities, pest-management strategies, and the development of appropriate IPM interventions.

> **Box 1—Transitional Crops and Use of Pesticides:** The transitional continuum extends from extensive to intensive production practices moving towards greater cropping, inputs (synthetic pesticides), and managerial intensity as a crop production becomes more oriented to the market economy. Specific crops at both African sites illustrate this basic concept of transition and the attendant increased use of pesticides. Sorghum at both sites is a subsistence crop, and its production is not associated with use of pesticides. Cowpea in Uganda is a true transition crop, having shifted over the last 15 years from being a subsistence crop to a locally important commercial crop. Its production is now associated with the use of synthetic pesticides. On the other end are green beans in Mali, an export crop, whose production relies heavily on use of synthetic pesticides.

Based on the above understanding of African agricultural systems as transitional systems, the IPM CRSP has since 1994 applied a participatory integrated pest-management (PIPM) approach to developing integrated pest-management systems for small-scale farmers in Mali and Uganda. The PIPM approach has relied on involving farmers in each step of the research process from problem identification and prioritization to farmer evaluation and redesign of on-farm trials. However, this transitional continuum, with varying levels of market integration and dependence on pesticides, has rendered sole reliance on either a purely tactical or an ecological approach inappropriate. What has emerged from the IPM CRSP experience in Africa is a synthesis of the two major approaches. The programmatic content of this synthesis combines development of a sustainable knowledge foundation based on both farmer and scientific knowledge and of programs that raise farmer knowledge and awareness of fundamental IPM concepts with development of pest management alternatives, including synthetic pesticide tactics when needed for priority pests and diseases. This approach to IPM program development and delivery has proven to be well adapted to the diverse transitional agricultural systems found at both the Mali and the Uganda research sites.

The remainder of this chapter will relate the experience of the IPM CRSP in sub-Saharan Africa to implement a participatory IPM program in diverse and transitional production systems. The focus is on the PIPM process, which, in addition to being participatory, has emphasized multi-disciplinary and multi-institutional collaboration and has made a concerted effort to involve women from the onset of the project.

The PIPM Process: Problem Identification and Prioritization, Crop-pest Monitoring, and Identification of Constraints to Adoption of IPM

The participatory process used at all IPM CRSP sites, and adapted to research sites in Africa to identify priority crops and pests and constraints on the adoption of IPM, consisted of a series of five activities. The process of project implementation at both sites began with a stakeholders' meeting, a participatory appraisal (PA) training session for scientists and field agents, and a PA with farmers and other stakeholders at selected research sites. In the African research sites on the IPM CRSP, field activities were initiated with a PA rather than a baseline survey as in several other sites. Activities were ordered in this way for several reasons. First, most scientists lacked knowledge of the cropping systems. The level of farmer knowledge of pests and diseases at the selected research sites was also undetermined. Scientist-farmer interaction was needed to create a common understanding for developing the research program and for establishing meaningful hypotheses for both formal baseline surveys and on-farm biological research. Finally, the PA made efficient use of the limited time of scientists from outside Africa. The CRSP modality involves international collaboration at limited, specific points of time, rather than the long-term posting of external scientists in national institutions as in some other projects.

The PA was followed by an initial baseline survey. Farmer perceptions of pest and disease priorities were then further documented through implementation of pest- and-disease monitoring programs. The outcome of these initial activities was the establishment of a farmer-centered IPM research (and to some extent, extension delivery) system and a solid foundation of knowledge on which to begin testing IPM component technologies.

Step 1: Stakeholders' Meetings

Stakeholders' meetings at both sites brought together lead administrators, a diverse and multi-disciplinary group of research scientists from the host countries and the United States, field extension personnel, and representatives from collaborating local NGOs. Priority research locations and crops were tentatively determined at these meetings. Additionally, the following institutional constraints on the adoption of IPM were identified at these meetings: 1) Weak linkages between pest management research scientists and farmers; 2) Poorly coordinated crop-protection efforts between research and extension; 3) Lack of proven crop-protection alternatives to synthetic pesticides; and 4) IPM research that did not take into account farmers' needs and knowledge.

Step 2: Participatory Appraisal (PA) Workshops

Participatory Appraisal (PA) workshops developed a shared understanding of IPM and the role of farmer participation in the research process, contributed to team-building, and helped refine a PIPM research and development strategy. At both sites, between 20 and 30 scientists and extension agents participated in designing the PAs by prioritizing important informational needs and selecting appropriate participatory methods. At the Mali PA workshop, participants were divided into small groups in which some participants demonstrated the use of one of the participatory methods while other participants role-played the part of farmers. At the Uganda Site PA workshop, PA methods were pre-tested with representatives from farmer NGO groups. Workshop evaluations at both sites indicated that the preferred sessions were those pertaining to participatory methods. The workshops ended by dividing the assembled groups into PA teams and making logistical arrangements for the conduct of the PAs.

Step 3: Participatory Appraisals (PA)

In both countries the initial activity for developing research priorities was the farmer-participatory assessment (PA). Farmer identification of priority problems is fundamental to the process of participatory agricultural research. Farmer-identified problems are used to determine research agendas, and it is this initial part of the process that differentiates participatory from other approaches to research-agenda planning. In addition to establishing farmer perceptions of priority pests and diseases, the PA site teams gained contextual insights into local systems and defined target audiences, as illustrated in Box 2 below. In Mali, multiple household-based production units form the context in which related family members meet their basic needs for food and cash. Each unit farms multiple fields differentiated by crops, management, and proximity to the village. Use of hired or part-time labor was minimal. In Uganda, households provide both labor inputs themselves and use non-household labor, with many farm households reporting the hiring of part-time labor or participation in labor exchanges.

Box 2— Contextual insights into causes of Striga parasitism derived from PAs in Mali and Uganda: 1) In the semi-arid zone of Mali, the preferred staple millet is grown near villages. These same fields are more likely to receive annual increments of manure and composted village waste materials than are "bush-fields" located some distance from villages. Farmers noted that millet in these fields tended to be "stronger" and survive Striga parasitism better than did distant bush fields; 2) In Uganda, farmers noted that Striga (Figure 4-1, page xix) had only become a problem with the loss of their cattle in recent times due to regional insecurity and cattle theft. Further questioning revealed that with the loss of cattle, crop acreage per farm family had declined, as had the practices of field rotation, fallowing, and the availability of animal manure. Farmers indicated that that fallowing and/or the addition of manure "weakens" Striga. They were also aware that continuous planting of sorghum in the same fields tended to build-up Striga populations. These two examples with different crops and different reasons for lack of fertility highlight a common principle, the effect of fertility on increasing the plants' capacity to withstand Striga parasitism. Finding the common principle first requires the contextual understanding.

Multi-disciplinary teams of scientists conducted the PAs with groups of farmers ranging in size from 12–70 at four selected research sites in each country. Following initial PA discussions, teams visited fields with farmers to observe and discuss problems and priorities. In Mali, participating farmers were from local villages and, owing to cultural norms, female farmers were interviewed separately. It was also necessary to spend some time initially with the village chief and elders to familiarize them with project objectives prior to meeting with younger farmers and groups of women. In Uganda, farmers were from active agricultural NGO groups, including a group with exclusive female membership. At both sites, a variety of participatory methods were used that produced priority rankings of crops and pests and diseases associated with these crops. These methods included the development of historical profiles that indicated changes in cropping systems and pests over the past 30 years. This method yielded valuable insights into how pest pressure had changed with transition in agricultural systems.

Step 4: Baseline Survey

With target audiences and crop priorities established by the PAs, baseline surveys were conducted to provide more in-depth and quantitative descriptions of local farmer characteristics, identify systemic relationships that affect pests and pest management at the farm level, and determine constraints to the adoption of IPM. In Mali, 171 farmers were surveyed. In Uganda, with support of the USAID Africa Bureau through the Integrated Pest Management Collaborative Network (ICN), 100 farmers were surveyed.

The baseline survey documented three aspects of pest management at both sites. First, in the minds of farmers and of scientists at the agricultural research institutions in both countries, pesticides dominated the pest-control agenda prior to the initiation of IPM CRSP activities. The PA and baseline surveys indicated that farmers in both countries preferred to use synthetic pesticides. In Uganda, 70% of the sampled farmers were already using pesticides either in the field or in post-harvest. Again in Uganda, 76 percent of the farmers growing cowpea, and 42 percent of the farmers growing groundnuts, reported using insecticides and all farmers growing tomatoes reported spraying their crop as many as 14 times per season. In Mali, despite governmental reductions in price subsidies for pesticides, 60 percent of the farmers at one research site were using pesticides on millet. A recent survey of green bean growers in Mali indicated that all surveyed growers were using pesticides. At both sites, pesticide use is associated more with cash-crop production; especially export crop production such as green bean production in the peri-urban areas surrounding Bamako.

Second, farmers at both sites were generally unfamiliar with other means to control pests, such as cultural and biological practices or the use of new resistant varieties. Knowledge of beneficial arthropods, except for spiders, was extremely limited. Research work and extension recommendations of pest management alternatives to pesticides were also limited when the project began.

Finally, incomplete farmer knowledge of pests and diseases, particularly of those that were less visible, suggested the need to introduce knowledge-based or soft technologies such as field pest- and disease-identification and monitoring programs in order to raise farmer knowledge of pests and diseases and involve them in the research process. Farmers in Mali lacked specific terms to differentiate different types of larvae, and used one term for all larvae. In Uganda, farmers were unaware of less visible insects such as the bean fly and thrips, and that insects (aphids) could vector diseases (groundnut rosette virus).

Box 3—Factors that may affect adoption of IPM: Farmers at both sites perceive labor and pests (insects, diseases, and weeds) as the two most important constraints on agricultural production. In Uganda, this perception is validated by farmers' demonstrated willingness to hire labor and to purchase pesticides and in Mali by the correlation between perceptions of wealth and access to animal traction. The implication for IPM is that pest-management technologies that result in increased labor demand may be resisted, and that technologies that demonstrate pest resistance or control without increasing the demand for labor may be rapidly adopted. This same phenomenon would also explain the attraction of synthetic pesticides: they are easy to use, reduce labor inputs, and have a visible impact on yield (Among the Teso in Uganda, the word for fertilizer and pesticides is the same).

Step 5: Follow-up pest-surveillance and monitoring programs

Knowledge of the biology of pests and diseases is vital to making informed pest-management decisions and is one of the underlying tenets of strategic IPM.

At both sites, the PA and the baseline surveys suggested that follow-on programs of pest surveillance and monitoring were needed to quantify the incidence and importance of pests and diseases noted by famers; to help farmers identify less obvious crop pests and diseases and understand indirect mechanisms of damage; and to introduce basic IPM principles and techniques of field scouting. In Mali, four field agents were trained to install and monitor light traps with farmers, as well as use pitfall traps and conduct periodic sweep-nettings to monitor insect populations. In Uganda, a farmer-implemented crop-pest monitoring system was deployed where crop-protection extension agents collaborated with an NGO field coordinator who then worked with five contact farmers to sample fields, record species and infestation levels, and document losses.

In both Mali and Uganda, pest- and disease-monitoring activities yielded important information. For example in Mali, monitoring activities documented fluctuations of pest infestations. Blister beetle infestations, an

Figure 4-2. Woman watering green beans in Mali.

important priority during the 1994 PA, were lower than that of a new species of Scarab beetle, *Rhinyptia infuscata*, detected during the 1995 field trials. In Uganda, monitoring activities altered farmer perceptions of pest priorities and identified new pests and disease species. During the PA farmers ranked the mole rat and monkeys as very important pests on maize, beans, and groundnuts. Subsequent field-monitoring activities provided evidence that these vertebrate pests were not nearly as important as farmers had perceived. The bean fly (*Ophiomyia* sp.), formerly unknown to farmers, was identified as the most important yield-reducing pest on beans. A survey of maize pests and diseases indicated that gray leaf spot (*Cercospora zeae-maydis*) was a seasonally important foliar disease and that termites (*Macrotermes*) were causing significant stand losses. Continuation of field-monitoring activities has identified other pests and diseases, including the first identification of a relatively recent arrival of a significant new pest on groundnuts, the groundnut leafminer (*Aproarema modicella* Deventer), and the periodic appearance of yellow blister disease, *Synchytrium dolichi* (Cooke), on cowpea. More recently in Uganda, thrips (*Thrips palmi* Karny, *Frankliniella schultzie* Trybom, *Scirtothrips dorsalis* Hood) were determined by scientists to be important new pests on cowpea and groundnuts.

Other benefits of crop monitoring activities included having a few farmers learn to "scout" fields, identify pests and beneficial insects, recognize the seasonal importance of various pests, and implement decision-making processes such as sampling and pest thresholds. Additionally, integrating farmers into the research process empowered them by including them as decision makers and knowledge producers. In Mali, after seeing how light traps attracted insects when used as a monitoring tool, farmers proposed testing these traps as a control tool. A trial in which pest insect populations were monitored at increasing increments of distance from the trap supported their observation quantitatively. In Uganda, one farmer, having observed that increased plant population was effective in repelling aphids on groundnut, proposed the same treatment for cowpea, with equally observable results. At both sites, farmers having observed initial IPM CRSP post-harvest trials began experimenting with different locally available materials.

From a research management perspective, integrating farmers into the research process also proved to be efficient. It not only reduced the logistical costs of scientists in the field, but also enabled more sites to be covered. In Mali, semi-arid research sites were 5-7 hours from where scientists were based. Having farmers work with field agents was the only way research could be effectively conducted under such conditions. Moreover, in most cases, farmers viewed research as a new opportunity, not as an added burden transferred to them. As one farmer in Uganda stated, "It [pest identification and monitoring] helps me enjoy my farming." In Mali, farmers considered having an on-farm trial to be an honor, a recognition of their capabilities as a farmer and their leadership position in the village. Farmers have competed for trials, and nearby villages have asked to be included.

These five activities occurred during the first three years of project activities in Mali and Uganda. Pest and disease priorities combined with farmer pest-management knowledge (or lack of knowledge) determined and oriented the research agenda at both sites. IPM is knowledge- and management-intensive and, for farmers at relatively low levels of agricultural development, knowledge of pests and diseases and alternatives to pesticides cannot be assumed. Thus, knowledge enhancement and sharing between farmers and scientists became a consistent underlying theme of future activities.

Step 6: Process for Designing and Testing IPM Technology

The process of designing and testing IPM technologies began with the participatory method of triangulating knowledge from the PA, baseline survey, and field-pest-monitoring activities to define, confirm, and refine crop, insect, disease, weed, and other pest priorities. These farmer-demand-driven priorities determined the scope of the research agenda. The next step in the PIPM process was to address site-specific priorities by developing, testing, or validating alternative insect-, disease-, and weed-management practices. Component practices suggested by farmers, scientists, and extension agents included pest monitoring and field scouting, development of economic thresholds, cultural practices, host plant resistance, biological control, and reduced, timely, and more effective applications of pesticides based on pest populations. These practices were then assembled and integrated into protocols for on-farm testing.

In Mali, where Striga on both sorghum and millet was ranked as a high priority, a set of trials was implemented to test various on-the-shelf technologies. For sorghum, four Striga-resistant varieties from IITA were tested in the field. For millet, a series of treatments was introduced including a Striga-resistant cowpea variety planted in alternated rows with farmers' millet variety, fertilization, and late weeding. The objective was to test for synergy of practices, as well as determine the best sequence of combinations. Farmer ranking of yield improvement and labor needs in a participatory evaluation of the trial matched the results of researchers' statistical analysis of plot yields, and led to farmer-researcher convergence on the best combination under farmer conditions. In 1999, an innovative approach to control Striga parasitism was begun using a herbicide 2,4-DB application to sorghum seed. First-season results demonstrated that this approach reduced the number and dry weight of Striga plants attacking sorghum by over 50 percent.

The Mali site concept of integrated treatments was replicated in Uganda for developing a Striga management strategy for sorghum. The trial components consisted of using a Striga-tolerant variety, fertilizer, and an indigenous plant suggested by farmers and known locally as Striga chaser (*Celosia argentia*). A laboratory investigation revealed that *C. argentia* induces suicidal germination and illustrated how scientific investigations in the laboratory can document farmer knowledge.

During the PA in Uganda, farmers indicated that Striga was less of a problem when cotton was used as a rotational crop. Thus a longer-term rotational trial was introduced to evaluate the effectiveness of rotating sorghum with trap crops — cotton and cowpea — to manage Striga. In Mali, maize and sorghum have been recommended to follow cotton in a three-year rotation to take advantage of residual effects of phosphate fertilization of cotton. While this recommendation was made with fertility management objectives, the Uganda results suggest that there may also be benefits for Striga management that have not been previously recognized. Results in one Africa site may thus validate practices in the other site.

This same approach, of testing and combining component technologies suggested by farmers and scientists, was also used for beans, groundnuts, and cowpea in Uganda. The treatments for bean fly consisted of an insecticidal seed dressing and the farmer practice of earthing-up around the stem. The most important problem on groundnuts, as perceived by farmers and verified by field-monitoring efforts, was groundnut rosette virus disease (RVD) that is vectored by the groundnut aphid (*Aphis craccivora* Koch). In Iganga district, farmers have reduced their groundnut production because of the severity of RVD. Early assessments of farmers' knowledge indicated that they were generally unaware of the role played by aphids in vectoring the disease. One of the component treatments tested was to increase plant population density. Higher plant density increases the inter-plant humidity and helps ward off aphid infestations. Other components consisted of early planting of groundnuts to avoid build-up of aphid populations; testing the newly developed RVD-resistant variety, Igola-1; and developing a reduced spray program consisting of 2-3 sprays beginning at 10 days after emergence and again at flowering.

The PA and survey activities repeatedly established that the most likely field crop in Eastern Uganda to be sprayed with chemical pesticides was cowpea. Cowpea has emerged as an important cash crop in this region with a lucrative export market in nearby Kenya. Over 70 percent of farmers growing cowpea apply pesticides; some farmers spray as often as 8 times per season. The major insect pests are pod-sucking bugs (*Riptortus* spp., *Nezara viridula*, *Acanthomia* spp., and *Anoplocnemis* sp.), *Maruca* sp., blister beetle (*Mylabris* spp.), aphid (*Aphis fabae*), and flower thrips (*Megalurothrips sjostedti*). The most important disease has been found to be cowpea mosaic virus (*Sphaceloma* sp.), although yellow blister disease (*Synchytrium dolichi* Cooke) is periodically devastating. Another constraint contributing to insect and disease problems is a lack of improved cowpea germplasm.

In response to these constraints, IPM CRSP plant pathologists are screening new germplasm provided by IITA for resistance to major cowpea diseases. Multiple-year on-farm testing has succeeded in assembling two IPM packages for cowpea that integrate well-timed insecticide spray applications (once each at budding, flowering, and podding) with cultural practices including early planting, manipulated plant densities, and/or cowpea/sorghum intercrop. These packages have been found to be effective in reducing insect pests on cowpea and increasing grain yield by over 90 percent.

Although most farmers were initially unaware of thrips, continuing field-pest-monitoring activities have determined them to be an important insect pest of cowpea. The relationship between thrip population density and cowpea grain yield loss has been found to be roughly linear and negative, and the economic injury level (EIL) for thrips (*Megalurothrips sjostedti*) has been established at 12 thrips per flower.

Green bean production in Mali is for export, and peri-urban producers were found using pesticides up to 4-5 times per growing season. The most important pests for green beans were determined to be thrips, whitefly (*Bemisia tabaci*), pod borers, and soil-borne diseases. An integrated package of technologies for green beans has been developed; it consists of neem applied twice at critical stages, use of yellow and blue traps, and soap applied when necessary as indicated by scouting.

Step 7: Participatory Technology Assessment

Another component of the process for designing and testing IPM technologies was to have farmers evaluate or assess trial technologies following on-farm trial implementation.

A participatory assessment of on-farm integrated-pest-management trials was conducted at research sites in Mali in 1996. Farmers found integrated Striga management and Striga-resistant sorghum varieties as being moderately effective in controlling Striga. For Striga management, farmers ranked their own practices lowest in yield but also lowest in labor requirements; conversely, they considered the combination of all Striga management practices together to be the most productive but also the most labor-demanding. Fertilization was seen as being less labor demanding and more effective in contributing to yield, as compared to late weeding. This latter finding was supported by agronomic results and led to modifications in trial treatments for the following two years. Late weeding was dropped from the combinations tested, while alternative intercropping arrangements of millet and the Striga-resistant cowpea were introduced to find the best combination that met the three farmer criteria of maintaining grain yield, avoiding excessive labor increase, and controlling Striga. Thus, farmer participation resulted in more relevant research.

In Uganda, farmer assessment of technologies was pursued using two mechanisms. First, an evaluation of IPM technologies was conducted with farmers participating in the on-farm trials. Second, a farmer field-day was hosted at research sites where farmers presented the various trial treatments, followed by group meetings to discuss the trials. The farmer field-days were considered more successful because they reached a wider audience and empowered local farmers. The group discussions also proved invaluable in altering some trial treatments.

In each of the IPM programs, basic data were collected by the biological scientists for every experiment so that partial budgeting could be completed to assess economic profitability. In many cases the IPM strategy that gave the highest yield did not result in the highest profits for reasons such as labor costs or seed costs. Therefore interaction among biological scientists and social scientists, including economists who helped design the instructions for data collection and for calculation of benefits, was important for developing recommendations for farmers.

The Role of Women in Pest Management and Implications for IPM

In recognition of women's critical contribution to agricultural production and food security in sub-Saharan Africa, it was important that special efforts be made to include women in each step of the participatory IPM research process. For example, during the PAs, different strategies were used at the research sites to ensure women's participation. In Mali, during initial legitimization meetings with village chiefs, requests were made to meet and conduct PAs with women. In all but one case, this request was granted. However, in one village the request was refused and another research site was selected. This particular village consisted of immigrant farmers with a division of male and female labor roles different from the dominant ethnic group, where women's roles in agricultural production are higher. This refusal to include women shows the importance of understanding interactions between cultural norms and agriculture. In Uganda, an NGO farmers' association with exclusive female membership was selected for collaboration, and women made up at least half of the attendees at other PA meetings. At both sites, an equal number of women were selected for extension and surveying activities. In Mali, the lead host-country scientist is a woman, as are two of three collaborating plant pathologists and the collaborating agronomist. Half of all on-farm research trials in Uganda were conducted with women.

Women play an important role in pest management decision-making in both countries, particularly for crops grown primarily as food rather than as cash crops. As an example, women at one research site in Uganda noted that cowpea used to be a woman's crop and sorghum was a man's crop. However, cowpea has become an important cash crop and is now considered a man's crop, and sorghum, whose value as a cash crop has declined, is now considered to be a woman's crop. Farmers at both sites, both male and female, perceive that labor and pests are the two most important constraints on agricultural production. Although division of agricultural labor by gender is a complex phenomenon in Africa, women's labor contribution to weeding at both sites is paramount. Thus, IPM practices that can reduce weed pressure would appear to directly benefit women. Men also appear to have greater access to land and production-enhancing inputs including fertilizers and pesticides. At research sites in Mali, significantly more manure and pesticides were reported being used on men's individual fields than on women's fields. In Uganda, baseline data indicate that men are more likely to be purchasing and applying pesticides; however, women were as likely as men to have pesticides used on their fields, but the women themselves were not applying the pesticides. Correspondingly, women were more aware of the hazards or the potential for negative impacts on human health from pesticide use, perhaps explaining why men are more likely to be applying the pesticides.

An assessment of project impacts in Uganda indicates that female participation (number and frequency of female participants) has been higher than male participation, perhaps because of the project's targeting of women, a greater concern by women for improving their agricultural production, or both. At both the Mali and Uganda sites, it is expected that in the future, as women through agricultural production participate more in the commercial economy, their use of synthetic pesticides will increase unless an alternative strategic approach guides the development of IPM technologies. Thus there is a continuing need to ensure the participation of women in the design and management of on-farm trials to aid their contribution to household food security.

Materials Development and Technology Transfer

The participatory approach has contributed significantly to technology development, but it needs to be complemented by traditional extension techniques for disseminating new IPM technologies to a broader audience. Participatory research activities are limited to small groups of farmers, because of the intensity of research and the demands of maintaining consistent and continuous contact with them. Research quality can only be maintained within project budgetary parameters on a carefully targeted but relatively small scale. A second reason is that participatory programs are more demanding than conventional on-station approaches.

Both IPM CRSP sites in Africa have responded to this challenge by developing materials such as fact sheets, by hosting farmer field-days, and by implementing modified farmer IPM field-school programs. These programs have exposed groups of farmers to IPM concepts, allowed them to participate in the research process through the implementation of applied research trials, and promoted the notion of farmers learning from farmers. The Mali site has collaborated with FAO, the U.S. Peace Corps, and World Vision, and the Uganda site has collaborated with a USAID-supported project called the IDEA in the development of materials for distribution to extension organizations and farmers. The Mali site's intercropping association of millet with Striga-resistant cowpea and the sorghum variety Seguetanta has been extended to other zones in Mali by the Projet de Developpement Rural (PDR). In Uganda the bean-seed treatment and GRV-resistant variety are being disseminated to farmers by field contact agents employed by the IDEA project, and IPM packages for cowpea and groundnut through farmer field schools.

Figure 4-3. Women learning about IPM in Uganda.

Conclusions

Many of the direct contributions of farmer participation to the IPM CRSP research program have been mentioned. However, it is the process of participation that has made the most important contribution to program development. The participatory process has created the dynamic that comes from the systematic interplay of scientific (formal) and farmer (informal) knowledge systems. This dynamic has led to a creative synthesis of IPM approaches, created new knowledge, built social capital, and permitted adaptation to diverse and transitional agricultural systems found at research sites in Mali and Eastern Uganda.

An important test of the PIPM approach is whether it leads to the development of appropriate IPM technologies to meet the production needs of small-scale farmers. Although it is our perception that the process has worked — it has developed appropriate IPM strategies for small-scale farmers — the final proof will be farmer adoption, if not adaptation, of these strategies. We now have several studies attempting to assess realized adoption and its determinants. Preliminary findings from these studies appear to indicate that intensive exposure to IPM concepts and strategies through training programs such as farmer-field schools maximizes chances of adoption. Improved varieties may diffuse from farmer-to-farmer, but adoption of multiple and truly integrated-pest management systems may require more intensive training. This result appears to verify those who have labeled IPM as conceptually complex. Additionally, conceptual complexity complicates assessing adoption and the development of appropriate measurement and assessment tools. Does enhanced farmer knowledge of natural enemies, field scouting, or pest and disease identification constitute adoption? Measuring impacts such as reduced use of pesticides is a more complex task with farmers who do not keep records of pesticide purchases, rates, or application frequency.

Certain aspects of the PIPM approach were less successful, and some limitations emerged. Fostering inter-organizational collaboration, although desirable, often founders on sharing costs and responsibilities because organizations have differing guiding philosophies, goals, and objectives. Two methods the IPM CRSP has found useful for fostering inter-organizational collaboration have been germplasm exchanges and joint funding of graduate students. Graduate student programs foster collaboration at two levels: funding and advising.

The participatory process is demanding and requires researchers to consistently follow-up and follow-through with farmers to be successful.

Over the course of the program, some difficulties were encountered in consistently adhering to the participatory process. Many of these difficulties were precipitated by the need to remain within budgetary parameters and to produce rapid results. Budgetary limitations, particularly in the early days, impeded our efforts to include or add new disciplines and to attract and retain social scientists, specifically agricultural economists, and restricted our capacity to maintain more consistent linkages between research scientists and farmers at distant research sites. However, the identification of dependable and field-savvy extension agents and graduate students helped to minimize some of these logistical constraints.

Additionally, the push and pull between the need to publish research findings and the need to have on-farm impacts may not always be compatible. There may be some inconsistency between implementing a fully participatory research program and the need to produce rapid research results. In our rush to obtain replicable results, we missed some opportunities to return information to farmers. On-farm trial designs were often too complex and plot sizes too small to be easily understood by farmers. Droughts, el-Nino rains, lack of vehicles, and the range of on-farm issues that can impede timely trial implementation and monitoring all conspired at one time or another to interfere with data collection. A general lesson learned was that potential short-term gains from attempts to modify or attenuate the participatory process usually were not realized and resulted in delaying the research process and the generation of on-farm impacts.

Finally, the demands of technology development and testing coupled with full farmer participation initially limited efforts to diffuse and expand the number of farmers reached by the program and may therefore suggest a two-stage approach. The first stage would focus on technology development and the second stage would focus on disseminating technologies to a broader audience and adapting these technologies to site-specific locations. Although the critique will be that this resembles the traditional system of technology transfer, the key difference would be maintaining farmer participation throughout the process. It remains to be demonstrated whether participatory technology generation and dissemination can be conducted simultaneously and effectively and on a large scale unless abundant financial resources are available (Quizon et al., 2001). Thus, a distinct technology-dissemination component, albeit with farmer participation, may be required to scale-up and reach a larger audience. Additional thoughts on this point are presented in Chapters 8 and 9.

References

Bentley, J.W., and K.L. Andrews. 1991. Pests, peasants, and publications: Anthropological and entomological views of an integrated pest management program for small-scale Honduran farmers. *Human Organization* 50(2): 113-123.

Cleaver, K.M. 1993. A strategy to develop agriculture in sub-Saharan Africa and a focus for the World Bank. Washington, D.C: World Bank Technical Paper: No. 203.

Dent, D. 1995. *Integrated Pest Management.* London: Chapman and Hall.

Ehler, L., and D. Bottrell. 2000. The illusion of integrated pest management. *Issues in Science and Technology*, Spring, pp. 61-64.

Kiss, A., and F. Meerman. 1991. Integrated Pest Management in African agriculture. Washington, D.C.: World Bank Technical Paper 142.

Mangan, J., and M. Mangan. 1998. A comparison of two IPM training strategies in China: The importance of concepts of the rice ecosystem for sustainable insect pest management. *Agriculture and Human Values* 15: 209-221.

Morse, S., and W. Buhler. 1997. *Integrated Pest Management: Ideals and Realities in Developing Countries.* Boulder, Co.: Lynne Rienner Publishers.

Norton, G., E. Rajotte, and V. Gapud. 1999. Participatory research in integrated pest management: Lessons from the IPM CRSP. *Agriculture and Human Values* 16: 431-439.

Pimbert, M. 1991. Designing Integrated Pest Management for sustainable and productive futures. Gatekeeper Series No. 29, International Institute for Environment and Development (IIED), pp. 3-16.

Quizon, J., G. Feder, and R. Murgai. 2001. Fiscal sustainability of agricultural Extension: The case of the farmer field school approach. Development Research Group. The World Bank: Washington, D.C.

Schwab, A., I. Jager, G. Stoll, and R. Gorgen. 1995. *Pesticides in Tropical Agriculture: Hazards and Alternatives.* Wurzburg, Germany: Margraf Verlag.

Yudelman, M., A. Ratta, and D. Nygaard. 1998. Pest management and food production: Looking to the future. Food, Agriculture, and the Environment. Discussion Paper 25: International Food Policy Research Institute (IFPRI). Washington, D.C., pp. 34-39.

— 5 —
Developing IPM Packages in Latin America

Jeffrey Alwang, Stephen C. Weller, Guillermo E. Sánchez, Luis Calderon, C. Richard Edwards, Sarah Hamilton, Roger Williams, Mike Ellis, Carmen Suarez, Victor Barrera, Charles Crissman, and George W. Norton

Introduction

IPM programs have expanded in both central and South America over the past ten years. The South-American program has been focused in the Andean region, especially on potatoes and fruits, and has been led in part by the IPM CRSP, the International Potato Center (CIP), and The International Center for Tropical Agriculture (CIAT) in conjunction with national agricultural research systems (NARS). Central American IPM has focused on non-traditional agricultural export crops, especially fruits and vegetables. The IPM CRSP has centered its regional IPM programs in Ecuador and Guatemala. In this chapter, we turn our attention first to the Ecuador program and then to Guatemala.

South America IPM Program in Ecuador

Ecuador, like much of the Latin-American region, is distinguished by its remarkable physical and ecological diversity. The varied topography of the heavily mountainous country includes coastal plains, highly sloped middle elevation tropical forests, high elevations with steep slopes, and tropical rainforests with Amazon tributaries. Like its topography, Ecuadorian agriculture is characterized by diversity; crops and farming systems vary even within well-defined agro-ecosystems, as do agricultural pests and means of managing pests. As a result of this diversity, Ecuador is particularly suited for the participatory IPM approach, as pest problems are localized and solutions must be tailored to meet local needs.

Ecuador is classified as a lower middle-income country with a PPP GNP of $2,600 in 1999 (World Bank, 2001). The country relies on agriculture for approximately 17% of its GDP, with close to the same percentage derived from oil and natural gas, Ecuador's chief exports. Major crops include banana, flowers, sugar cane, rice, maize, plantain, and potatoes. The latter two crops represent important food staples; potatoes are widely consumed by lower and middle-income groups in the highlands and plantains in the coastal regions. Plantains are increasingly being exported, particularly to Colombia and Central America. Among agricultural exports, bananas reign supreme, accounting for about 85% of the close to $1 billion in agricultural exports in 2001. Plantains are the third most important agricultural export, representing about $18 million in exports in 2001. Plantain production and exports represent a significant source of agricultural growth, particularly in the coastal regions of the country.

Potato production is widespread[1] in the highlands of Ecuador, where more than 44% of the country's nearly 13 million people live. Plantains are grown in areas along the coast and at mixed elevations throughout the tropical regions of the country; FAO estimates that Ecuador has approximately 70,000 hectares of plantains. In coastal areas, plantains represent a major food staple (Figure 5-1).

Ecuador's public agricultural research is principally conducted by the Instituto Nacional Autónomo de Investigaciones Agropecuarias (INIAP), a semi-autonomous agency that was formed to make agricultural research in Ecuador more productive by introducing market-based incentives. INIAP's semi-autonomous structure is suited for project-based support because it is flexible and responsive to client needs. These attributes also make it an appropriate partner for participatory IPM. In the early 1990s, INIAP underwent an administrative reorganization that positioned the institution for participatory IPM work. Its Plant Protection Department adopted an integrated approach to basic and applied research across regions, commodities, and disciplines. The Department had begun training scientists to participate in a country-wide IPM network and had also undertaken a comprehensive assessment of main pests and diseases of vegetables and fruits. As of 1994, the institution was poised to begin intensive work on IPM, but lacked funding and scientific expertise. The IPM CRSP was invited to help fill the void. Although Ecuador was listed as a secondary site in the original IPM CRSP proposal, funding for research in that country began in 1997/98.

[1] Approximately 60,000 hectares of potatoes are grown in Ecuador (FAO).

Figure 5-1. Plantain renovation experiment.

Identifying and Prioritizing IPM Problems and Systems

IPM priorities in Ecuador were determined through a combination of stakeholder meetings, participatory appraisals, an assessment of current information about pests and pest-control measures, crop-pest monitoring, an *ex-ante* analysis of potential impacts of IPM, and analysis of institutional strengths. INIAP and U.S. scientists conducted a series of stakeholder meetings attended by representatives from CIP, PROEXANT (a non-traditional agricultural export assistance firm), CARE, USAID, the Ministry of the Environment, and OIKOS (a local environmental NGO), among other organizations. These meetings, in combination with participatory appraisals, were used to identify serious pest problems in potato, fruit, and plantain cultivation. In the case of potatoes, farmers in one of the main potato-growing regions (Carchi, along the northern border with Colombia) were concerned that pervasive pest problems were inducing over-applications of pesticides and causing reduced yields or storage losses, leading to lower competitiveness of Carchi potatoes and adverse health effects. In plantain regions around El Carmen, growers identified a dearth of pest-related information and were reliant on recommendations from banana exporters. Potatoes and plantain were logical crops for a PIPM focus, due to these problems, the economic importance of the two crops, and a varied institutional base on which to build. This base led the scientists to conclude that IPM in potato and plantain had a high probability of having important economic and institutional impacts.[2]

Highland Potatoes

In potatoes, several institutions had been engaged in basic and applied research and, to a limited extent, technology transfer, both at the Santa Catalina experiment station near Quito and on-site in Carchi. These include the International Potato Center (CIP), INIAP's own National Potato Program (FORTIPAPA), and the Soils CRSP among others. In addition, Eco-Salud, a health-oriented non-governmental organization, had been conducting research on the health and environmental effects of high-input potato production. The research base provided by these entities, their integration into the community, and the possibility for outreach of results[3] created opportunities in potato IPM that did not exist for other crops.

Potatoes are grown in the Carchi region in a mixed farming system where potatoes are primarily rotated with pasture for livestock.[4] Households in the region also produce beans, barley, corn, and a variety of horticultural products. Holding sizes vary, with a mean holding of 2-3 hectares; the typical household produces 0.75-1 hectare of potatoes. Typical holdings are steeply sloped but fertile lands; approximately 33 percent of the land in the area is irrigated. Potato yields are approximately 15,000 kg per hectare, compared to 11,400 nationally (FAO, 2003). The area's economy is heavily dependent on agriculture; about 45 percent of the workforce is comprised of agricultural day laborers, and another 33 percent is described as self-employed (most in agriculture). The largest sources of family incomes are milk and potato sales. Sixty-three percent of the population has received primary education of some type, while 23% have some secondary education. More than 10 percent has had no formal education at all. Farming is a way of life in the area, and men, women, and children share production

---

[2] The IPM CRSP also worked in Ecuador at middle elevations on pest problems of several Andean Fruits and a mixed coffee-plantain system in a fragile ecosystem.

[3] Ecuador has no formal public agricultural extension service. As a result, INIAP is continually searching for innovative means of extending research results.

[4] Much of the information in this section comes from the IPM CRSP baseline survey for potatoes, conducted in 1999.

duties. Women, in particular, are actively engaged in farm decisions and participate in pesticide use and storage.

The major potato pests identified during the IPM prioritization process were late blight (*Phytophthora infestans*) (see Figure 5-2, page xix), Andean weevil (*Premnotrypes vorax*), and Central American tuber moth (*Tecia solanivora*). The former is a worldwide limiting factor in potato production and, prior to the CRSP, the farmers' primary means of control was intensive field spraying of fungicides, particularly Cymoxanil and Mancozeb. Weevils and tuber moths were also controlled through heavy spraying, with Carbofuran, a highly toxic member of the Carbamate group, being the principal agent. The tuber moth also attacks potatoes in storage. Lack of information on pesticide applications led to widespread mixing of as many as 12 chemical agents into "cocktails." The average farmer made 8 chemical applications in a single growing season with 2-3 insecticides or fungicides in each. In fact, analysis of the base-line data showed that only about 25 percent of farmers applied pesticides in "good" combinations and far less than 50 percent used the appropriate dosage. Women, who in recent years have taken increasing responsibility for management of the farm's affairs, were particularly concerned about the misuse of pesticides, its impacts on farm income and on human health.

The potato IPM strategy

The research base provided by CIP and FORTIPAPA facilitated the CRSP beginning work on pest-management methods at several levels including experiments under greenhouse conditions, trials in farmer fields, and participatory experimentation in farmer field schools. For example, CIP/FORTIPAPA had identified a number of potato clones with horizontal resistance to Late Blight. These clones had not been examined in on-farm conditions nor for farmer acceptance. In fact, farmers in Ecuador are particularly concerned with factors such as coloration, tuber appearance and taste, and consumer acceptance. In the first years of the IPM CRSP, a project that tested these clones in multidisciplinary on-farm pest-management experiments identified three varieties that comprised an integral part of the IPM package for potatoes (Figure 5-3). Late blight damage is controlled in the package through resistance and more limited fungicide applications.

Laboratory experiments sponsored by complementary projects had measured the effectiveness of several low-toxicity products in controlling the Andean potato weevil. The CRSP, building on these findings, immediately

Figure 5-3. IPM CRSP collaborators in Ecuador viewing potato experiment.

began field trials of eight promising products and identified a chitin inhibitor, Triflumuron, as the best means of insect control. During evaluation of these products, their effectiveness in traps, bait plants, and foliar application was tested and the results used to formulate a trap/foliar application strategy that would form the basis of the IPM package for control of the weevil. During subsequent years of the project, foliar application methods were refined as a part of a farmer field school participatory learning experience, and chemical application on the lower leaves of the plant on alternating rows of potatoes was determined to be the best method of control. These recommendations were immediately incorporated into the IPM package.

In the case of the Central-American tuber moth, farmers immediately identified damage caused by the pest during storage of tubers as its most serious impact. As a result of this damage, farmers relied on heavy applications of pesticides in storage areas, areas often involving storage of other foods and household implements. The CRSP began working in collaboration with French researchers and staff at the Catholic University of Quito to study the effectiveness of biological control methods, particularly Bacolovirus. Because of the existing research base, CRSP researchers were able to begin their investigations in farmer storage facilities. Researchers discovered that more virulent strains of Bacolovirus were needed for effective control, but existing strains, in combination with low-level applications of Carbaryl, were deemed to be an effective temporary means of controlling the pest. As a part of this participatory trial, researchers were closely engaged in observing potato storage techniques; observation of the techniques permitted researchers to propose and test more effective means of tuber storage including simple low-technology storage silos. These silos also form part of the recommended IPM package (Figure 5-4).

Socioeconomics in potato production

Socioeconomic concerns were completely integrated into the potato research, starting with identification of stakeholders, interactions with producers and experts to understand key problems, and evaluation of solutions to pest problems. This integration enabled the researchers to move quickly toward socially acceptable solutions and facilitated efforts at out-

Figure 5-4. Seed potato storage in Ecuador.

reach and evaluation of IPM program impacts. In particular, the baseline survey, whose analysis was well underway at the start of the second year of the project, helped tailor subsequent technical and socioeconomic research. Concerns expressed by producers about taste, coloring, and other potato properties helped guide the selection of resistant clones. The key concern was market acceptance of the product — an economic problem. Problems with chemical and potato storage led researchers to investigate alternative storage systems. Information about surplus labor on small-scale potato farms allowed researchers to investigate labor-intensive practices such as scouting, trapping, and use of labor-intensive methods to remove leafminers from fields.

Gender analysis comprised a major portion of the socioeconomic work and provided important insights into the research. During the process of creating the baseline, women identified health problems as their primary concern relative to pesticide use on potatoes; subsequent research was designed to minimize adverse health impacts. The baseline survey examined decision-making within the family and showed that women and men jointly make many decisions about production systems and field operations and that women make important decisions related to pesticide storage and its proximity to other household activities. As a result of these findings, female producers were brought into the farmer field schools and played a key role in sensitizing the men about adverse health effects of improper pesticide use. Women's groups were also formed as a means of educating and empowering women to influence important health-related and other household decisions.

Research and outreach were informed by economic analysis. All the technical activities collected information on costs of alternative practices; these costs, combined with information about impacts on yields, helped identify economically optimal pest-control methods. The farmer field schools used this information to identify alternative IPM packages, and the schools themselves collected further information on costs of different practices during field trials. One result of combining experiment-field-based economic information with that information gained through field schools was the knowledge that costs and yields based on a small experimental plot could not just be expanded to produce yield estimates for larger fields. In fact, experiment-field-based data generally tend to understate costs and overstate yields when projected to full field estimates.

By the end of the second year of the project, field-school participants in Carchi knew that the net benefits from an IPM package involving a late blight-resistant variety (INIAP-FRIPAPA99), traps and limited leaf spraying for Andean weevil, monitoring and limited spraying for the tuber moth, and other low-input controls, were associated with a $643 per hectare net benefit when compared to local practices. Other field-school experiences showed similar magnitudes of benefits. An impact analysis of a single program (late blight resistance) in the Northern Region showed a net present value of research of almost $34 million.

IPM for Plantain

Two main plantain production systems exist in Ecuador: monoculture plantain predominating in coastal areas and mixed cacao-coffee-plantain systems at low and intermediate elevations. The former system typically involves medium to large holdings averaging 40-70 hectares, depending on location, while the latter is exercised on smaller holdings, varying anywhere from 5-25 hectares in size. The IPM CRSP decided to focus on monoculture plantain for several reasons. First, pest problems appeared to be most limiting in the monoculture system. Second, the information base for integrated control of plantain pests was minimal, and researchers decided to address the plantain problems first before tackling them in a mixed system. Third, INIAP made an institutional decision to invest in plantain research and, because of limited human and financial resources, was unable to support too-diversified a research program.

The institutional base for plantain research was entirely different from that of potato, and IPM CRSP research on plantain began with a narrow information base. Unlike potato research, few domestic organizations were involved in plantain, the International Banana Research Center did not have an active program in plantain in Ecuador, and the Pichilingue Agricultural Experiment Station, where plantain research is centered, had limited human resources. The IPM CRSP has helped address each of these problems.

Common wisdom held that because of the botanical similarity[5] between plantain and banana, plantain pest problems were likely to be similar in nature to those of bananas, and control techniques practiced by banana producers would also be applicable to plantains. As multi-national corporations are key players in banana production and export, banana research is largely conducted by private firms; pest-control methods in bananas are based on export company recommendations. These recommendations, as adopted by plantain producers, became the base of comparison for IPM CRSP plantain research.

[5] Both belong to the Musaceae family.

The major plantain pests and diseases were identified through participatory appraisals conducted in the El Carmen region of Ecuador and subsequent pest-monitoring exercises sponsored by the CRSP. These problems are: Black Sigatoka (*Mycosphaerella fijiensis*), black weevils (*Cosmopolites sordidus*), and parasitic nematodes (*Helicotylenchus* and *Meloidogyne*). Black Sigatoka is a fungal disease, and is recognized as a severe problem that, due to poor management practices, had become endemic to the region. Banana export companies' control strategies for Black Sigatoka include regular heavy spraying of a fungicide (including varying amounts of Benomil, Tilt, and Calixin) mixed with agricultural oil combined with weekly sanitary leaf pruning. Black weevils (also known as the banana root borer) were also recognized by producers as a limiting pest, and subsequent research indicated that as much as 30 percent of the yield loss in plantain was caused by this pest. The weevil causes most damage during its larval stage when it feeds on the plant's corms. Local recommendations for black weevil control included aggressive spraying with highly toxic insecticides such as Furadan (Carbofuran) and Lorsban (Clorpiriphos). Nematodes were not recognized by local farmers as a pest problem, and were only identified by the CRSP as such after the project's second year.

Integrated management of plantain involved more than management of its pests. The monoculture producing region is comprised of an abundance of plantations that had suffered for 20 or more years of poor management. Poor management was manifest in improperly spaced and cleaned trees, overgrowth of weeds, and resultant low yields. Improvement of production systems required decisions about complete renovation or rehabilitation of these poorly managed plantations. Renovation involves removal of the entire plantation stock and replanting, while rehabilitation involves selective removal and replanting. IPM CRSP researchers decided early in their plantain work to investigate the relative benefits of renovation versus rehabilitation; this decision required integration of economic analysis into the research from its very start. The monoculture plantain production system is also characterized by a high degree of involvement of women in farm decisions, and the CRSP decided to invest resources in understanding how decisions were made. This information will help inform design of an outreach program in plantain.

The research program

Plantain research combined laboratory experiments, experiment station plots, and participatory on-farm experiments. The former two were conducted at INIAP's Pichilingue Experiment Station, located outside Quevedo in the Coastal region of the country. The participatory on-farm experiments, focusing on creation of a plantain IPM package and evaluation of renovation versus rehabilitation of existing plantations, were conducted in El Carmen.[6] In addition to testing IPM packages and evaluating rehabilitation versus renovation management systems, on-farm experiments investigated alternative means of trapping black weevils and the impacts of IPM practices on black weevil populations. Laboratory experiments focused on testing and generating local strains of entomopathogenic fungi (*Beauveria bassiana*) to control black weevil. Surveys were also undertaken to determine the incidence and economic importance of nematodes in plantain. These surveys were taken throughout the "Plantain Belt," which encompasses some 45,000 hectares of monoculture plantain.

The on-farm experiments formed the core of the plantain research, and were designed to identify and test an IPM package, including recommendations about rehabilitation versus renovation, within 3 years of project inception. Plantain reproduction time, the rate of return from one generation to the next, varies depending on cultural practices, and the rate of return is a major determinant of economic viability of alternative management systems. Additional variables considered in the on-farm experiments were: disease index, youngest infected leaf, area under the disease progress curve (AUDPC), number of black weevils and nematodes present,[7] plantain yields, and costs of the treatments. Four treatments were evaluated (Table 5-1) for the rehabilitation trial; treatments were distributed in a randomized complete block design with four repetitions.

The large number of variables evaluated during the trials and the evolutionary nature of the research helped solidify our knowledge of the plantain insect/disease complex (see Figure 5-5, page xx). For instance, levels of infection with Black Sigatoka were relatively high on all treatments and fungicide applications originally appeared to be effective, when using the disease index variable, at controlling the disease (45-day intervals were no different than 30-day intervals). However, analysis of AUDPC values showed that except in years with extremely high rainfall and favorable

---

[6] Formally, the participatory on-farm experiments were conducted at the "Santa Marianita" farm, owned by the Carranza family. It is located in San Augustin Parish, Manabi Province, 250 meters above sea level. The average temperature is 24°C. with 2900 millimeters of annual rain.

[7] For nematodes, the severity of diseased roots was measured using a locally developed scale.

Table 5-1. Description of on-farm participatory experiments, monoculture plantain

Treatment	Description
Export Company	Weekly sanitary leaf pruning (removal of entire leaf where any necrosand is found); fungicide application with Benomil (280g/ha), Tilt (4l/ha) and Calixin (.6l/ha)—2 successive sprays monthly; trapping of Black Weevils; weed control with Glyphosate; 3-4 daughter plants per reproduction unit
IPM with fungicide	Leaf surgery (removal of only the necrosand area) every 15 days and management of pruned leaves on the ground to avoid dispersal of inoculums; 45-day interval for same fungicide application as under Export Company recommendations; monthly trapping of Black Weevils using V-traps with newly harvested pseudostems used as bait; manual weed control combined with natural cover (*Geophila macropoda*); fertilizer following soil analysis; selection and ordering of productive units to cover empty areas (rehabilitation trial); no more than 3 daughter plants.
IPM without fungicide	Same as above, without fungicide applications
Farmer's practice	Light weeding twice yearly; annual cleaning of plants (removal of dead leaves).

disease conditions, differences across treatments were not statistically significant. The full analysis, including evaluation of costs and values of plantain production, showed that fungicide applications were less successful without complementary management practices and that, with good IPM practices, fungicides were not economically recommended except in years of extremely heavy disease incidence.

Experience during the first two years of the on-farm trials helped identify nematodes as a significant source of damage to plantain plants, but black weevils were shown to be the most damaging plantain pest. Measures to control black weevil damage became a critical focus of IPM CRSP researchers because trapping methods were not yielding promising results. In fact, black weevil infestation levels did not respond to any of the on-farm treatments using a variety of traps and baits. As a result of these findings, activities were initiated to investigate alternative biological controls for black weevils, to understand black weevil behavior under natural conditions, and to develop alternative trapping systems, including use of synthetic attractants. Experimental observation of black weevil behavior at different times of day, seasons, and points in the life cycle helped researchers design and evaluate improved trapping systems. Experiments with different attractants led to improved trapping methods, and an alternative trapping system is now being evaluated in on-farm trials.

Socioeconomics in plantain production

Little was known at the start of the plantain activities of conditions in producing households, decision-making within the households, or the degree of knowledge about plantain pests and pesticides. In contrast to the potato-growing region, few studies of social conditions in plantain-producing regions had been conducted. In addition to the relative absence of this information, CRSP researchers were interested in investigating alternative methods of dissemination of IPM technologies. Farmer field schools had been the principal source of dissemination and capacity building in potato-producing areas, but plantain researchers were skeptical of direct application of the field school methodology in the plantain region. Field schools have been shown by others to be a relatively expensive means of reaching participants (Quizon, Feder, and Mugai)[8]. Field schools also would be a less appropriate means of eliciting participation of neighboring farmers because of landscape differences between potato- and plantain-producing regions and because production units are so large in monoculture plantain regions. IPM CRSP researchers decided to consolidate the baseline survey of socioeconomic conditions in plantain regions with an exercise that gathers information about how knowledge is transmitted among plantain producers. This process helped lower costs of information gathering and ensure

[8] In fact, in the potato region, more than $50,000 has been spent using IPM CRSP funds in field schools with a total of 302 participants. Although the money was also used to promote field days with roughly 1700 person-day participants and field-school participants are encouraged to share their experiences with neighbors, this represents an expensive means of transmitting knowledge. The $50,000 figure understates the true cost because FAO and INIAP also provided counterpart funds for the field schools.

that technology dissemination measures were optimally designed with respect to cost, message content, and audience.

Economic evaluation of IPM in plantain began with partial budget analysis that was built directly into the research. This analysis helped establish the economic viability of alternative treatments and the importance of evaluating the "rate of return" of plantain daughters under the different treatments. Budget information, along with an assessment of likely adoption rates from interviews with field staff, is being used to produce an assessment of impacts of the research program. Plantain producers are frequent employers of hired labor, and IPM measures are expected to have strong off-farm impacts through labor market linkages. The assessment of economic impacts of IPM in plantain thus had to consider these off-farm impacts. Analysts were also interested in impacts of such research on the poor and lower-income groups. The impact assessment was tailored to examine the distribution of benefits and costs of IPM research among producers, consumers, and suppliers of labor.

Interdisciplinary Analysis and Institutional Linkages

The PIPM approach in Ecuador benefited from strong linkages to INIAP, the public agricultural research institution, but linkages with other institutions helped strengthen the overall program. Linkages with institutions such as CIP, Eco-salud, PROMSA (World Bank-supported agricultural technology and training project), and others deepened the financial resource pool and broadened and deepened the pool of expertise in support of project objectives. These linkages were especially important in facilitating and promoting interdisciplinary analysis. INIAP staff is almost universally trained in the biological sciences, and interactions with other institutions, whose expertise was often socio-economic in nature, helped build capacity for and appreciation of multidisciplinary analysis. Entry of the project into national policy dialogue together with the heavy emphasis on documenting impacts also helped create a bridge between the biological science and policy analysis.

The project also benefited from different degrees of institutional support at the start. As noted, the institutional base in potato-producing areas was substantial from the inception of the project. This base helped researchers quickly enter into validation and outreach activities. Plantain researchers had to build the institutional base, and were able to learn from the potato areas the importance of the institutions. Thus, INIBAP[9] was consulted early in the project and became intimately involved with scientists, local banana and plantain societies were used to promote and guide the on-farm experimental work, and private technical assistance providers have been engaged in the on-farm activities and in designing the outreach program. These institutions have helped build the technical base, have given the project legitimacy among producer groups, and have generally enhanced the project's impacts.

Central America IPM Program

The Central American IPM program has focused on non-traditional vegetable and fruit export (NTAE) crops that are important for the U.S. market. Fruit and vegetable exports from Guatemala have increased to more than $360 million (Sullivan et al., 1999). It is imperative that these exports be grown according to accepted practices to meet sanitary and phyto-sanitary standards of the importing country and ensure a safe food source. Production should also be economically viable with minimum adverse impacts on the environment and the health of producers. The overriding challenge faced by the NTAE sector in Central America is to produce crops that are accepted in the international marketplace. The main threats to this challenge have been high dependence on chemicals, lack of sustainable bio-rational practices, lack of institutionalized phyto-sanitary post-harvest standards, and lack of pre-inspection protocols for product pre-clearance and certification (Julian et al., 2000b).

The IPM CRSP has designed research and education programs focused on scientifically integrated crop management (ICM) production systems for NTAEs. The program is research-focused and interdisciplinary, relying on development of 'total system' management strategies to achieve greater sustainability and increased competitiveness in NTAEs. The research focuses on: (1) reduction of agricultural crop losses due to insect, disease, weed, and other pests while relying less on pesticides and chemicals; 2) production of high quality crops that are less susceptible to post-harvest degradation; (3) production and post-harvest practices to allow producers to meet established phyto-sanitary standards and regulatory compliances in export markets; (4) reduced damage to regional ecosystems; and (5) enhanced overall socio-economic conditions of producing families.

[9] The banana research CGIAR institution in Montpelier, France.

Program Establishment

The initial step in establishing the IPM CRSP program in Central America was to hold participatory appraisals with stakeholders in production, research, and marketing in order to identify the most important issues facing the NTAE sector. The PAs, held in Chilasco/Baja, Verapaz, and Chimaltenango in the Guatemalan highlands, lasted two weeks and were followed by a research planning workshop to identify research activities. These activities revolved around targeted crops, pests, production programs, post-harvest handling, and marketing. Stakeholder meetings were held in 1994 among representatives of the national agricultural research institution (ICTA), the University of Del Valle (UVG), CARE-Guatemala, AGEXPRONT (comprised of more than 300 producers and marketing groups), ALTERTEC (which works on biological control and organic methods), AGRILAB (which serves the non-traditional crop sector with disease management and soils testing), the PanAmerican Agricultural School at Zamorano, Honduras, FLASCO (a resource institution on socioeconomic analysis), and the following U.S. institutions: Purdue University, Ohio State University, Virginia Tech, and the University of Georgia.

Following the PAs, a market assessment helped to identify windows of opportunity in U.S. markets, non-economic barriers to export trade expansion, needed policy revisions related to chemical registration and labeling, and market linkages from growers to retail buyers in the United States. IPM activities were prioritized around specific crops: broccoli, snowpeas, and brambles. Other crops including tomatoes, melons, papaya, mango, and starfruit were identified as important future IPM research targets. Crop-pest monitoring and baseline surveys helped measure the importance of pest, production, and handling problems.

The approach then shifted to connecting IPM research with a long list of major collaborators through establishment of a site committee. The research was designed to implement replicable and sustainable approaches to ICM focused on development of 'total system' management strategies. Crop-based research was conducted on farms, at research centers, in the greenhouse, and in laboratories. Socioeconomic research emphasized gender issues and farmer health. Strategic planning was used to identify mechanisms for transferring results to stakeholders. These mechanisms included training meetings, seminars, pilot production programs, and publications.

Results

Research found that the NTAE sector in Guatemala continues to enjoy a regional advantage in the production of horticultural crops targeted for sale in North America and Europe. Guatemala derives its economic advantage from an abundance of small-farm family labor, diversified microclimates for producing high quality counter-seasonal crops, and access to low-cost transportation infrastructure. Research suggested that future expansion of economically sustainable production in Guatemala would depend upon the industry's capacity to address increasingly important non-economic constraints to interregional trade, particularly more demanding food safety standards in the United States. The main non-economic constraints related to phyto-sanitary compliance and contamination from unapproved chemicals. A few examples illustrate some of the dimensions of these constraints.

Snowpea IPM

Snowpea programs illustrate how an initial interest in IPM can ripple through the entire NTAE sector, helping to achieve economic and environmental goals across multiple cropping systems. Guatemalan snowpea production has continually been harmed by insect and disease infestations leading to excessive reliance on chemical control measures. Leafminer (*Liriomyza huidobrensis*) became a major insect pest whose importance was magnified in 1995 when the USDA placed a Plant Protection Quarantine (PPQ) at U.S. ports-of-entry on all Guatemalan snowpea shipments due to fear that the species on Guatemalan snowpeas was an exotic pest in the United States (Sullivan et al., 1999). Research protocols were established to address the snowpea leafminer quarantine problem in March 1996. IPM CRSP research documented that the Guatemalan leafminer was not a species exotic to the United States, and consequently not a threat to U.S. producers. As a result, the quarantine was lifted (Sullivan et al., 1999). The process illustrates the need for proactive approaches to IPM.

The IPM CRSP recommended several testable strategies to reduce chemical residues on snowpeas and enhance product quality. The approach included non-chemical research regimes integrated into a holistic system. The main insect pests monitored were leafminers and thrips (*Frankliniella* spp.). Disease monitoring included Ascochyta (*Ascochyta pisi*), Fusarium wilt (*Fusarium oxysporum* f.sp. *pisi*), and powdery mildew (*Oidium* spp.).

In snowpeas and most other NTAE vegetables, the main pest-control strategy relied on the application of chemical pesticides using a 7- to 10-day calendar schedule. Few farmers in Guatemala were acquainted with IPM strategies, and most relied heavily on agrochemical distributors for pest-management information. IPM tactics included pest scouting for insects, pathogens, and weeds. Leafminer sampling occurred once a week to determine adult insect pressure, thresholds, and the need for pesticide applications. Sticky traps were used to reduce adult insect leafminer pressures and to determine adult insect thresholds. Row hilling was used to reduce adult leafminer reproductive capacities. Environmental Protection Agency (EPA)-approved pesticides were applied to the IPM plots. EPA-approved fungicides were applied sparingly, depending on environmental conditions. Grower-managed control plots followed traditional calendar pesticide application schedules at 7- to 10-day intervals regardless of insect pressure or growing conditions. Factors evaluated were snowpea export-quality yield, and leafminer larval populations 35, 65, and 90 days after planting.

ICM plots required only 3.7 pesticide applications, while traditional chemical-control plots required an average 10.4 pesticide applications (Sullivan et al., 2000). Insect populations and diseases were similar in ICM plots and control plots. In nine out of nine comparisons, ICM plots required fewer insecticide sprays; and in seven out of nine comparisons, the ICM plots had higher yields. These results translated into an average 61% reduction in pesticide use, and a 6% increase in average total yield. Product quality was found to be higher in the IPM plots, as measured by marketable yields at the shipping-point grading facilities. Product rejections at the shipping point averaged 6% less from the IPM plots. These results indicate that production and export of high quality crops are possible using IPM strategies.

Another example of leafminer control research in snowpeas investigated the potential of faba bean as a trap crop for leafminer (Sullivan et al., 2000). Trap crops are used to attract insect pests away from the primary income crop, thereby lowering insect pressure and reducing the need for chemical applications. Plots were managed using good agricultural practices without use of pesticides.

Results from greenhouse and field experiments showed that faba beans can be an effective trap crop for leafminers and can protect snowpeas from infestation. In preference studies in the greenhouse where leafminer females could choose between faba beans and snowpeas for egg laying, there was a significantly higher number ($P < 0.01$) of eggs laid on faba beans (5.41/gFW) than on snowpeas (0.12/gFW), and the number of emerging pupae from faba beans (2.14/g FW) was also greater than from snowpeas (0.25/gFW). Similar results were observed in field studies where the numbers of emerging larvae were higher in monoculture snowpeas than in snowpeas surrounded by faba beans. These results supported a hypothesis that faba beans are a preferred host for leafminer oviposition and can serve as a season-long trap crop. Currently over 30 percent of Guatemalan snowpea growers are using the faba bean trap crop in their ICM management programs in snowpeas (Sullivan et al., 2000). Other snowpea cultural and pest management experiments included intercropping, scouting, trap cropping, mobile trapping, minimum threshold pesticide applications, optimum crop cultural practices, and cultivar selection (Weller et al., 2002).

Tomato IPM

Another example of effective IPM research is in tomato. IPM research involved testing five approaches to reduce incidence of the whitefly-geminivirus complex and then validating these results in farmer fields. The tactics were: (1) seedbeds covered by a foamy cloth (anti-aphid covering), (2) tomato seedlings grown in newspaper transplant plugs, (3) use of sorghum barriers planted 45 days before tomato transplanting, (4) use of plastic sticky traps, and (5) rotational use of insecticides (rotating chemical groups) and sampling of whitefly populations. These treatments were compared to grower practices that included: (1) purchased tomato seedlings, (2) no sorghum barriers, (3) no sticky traps, and (4) programmed use of insecticides (without rotating chemical groups) and without sampling populations. Results showed that IPM production costs were $700/ha lower, profits were $1,700/ha greater, and pesticide sprays were reduced from more than 23 in grower plots to 13 in IPM plots (Weller et al., 2002).

Impact and Effect of IPM Research on Farmer and Industry Attitudes

The ICM strategy developed on the CRSP is applicable to multi-crop systems and is a critical element for achieving potential long-term sustainability in the NTAE production sectors of Central America. Programs must be transferred to the growers as a first step in the institutionalization of IPM. The transfer of IPM technology occurs mainly through field visits to research and validation plots. For example, between January 2000 and May 2001, there were 32 different extension activities, involving more than 1,000 participants (farmers and technicians from export companies, chemical companies, private and public organizations). These activities involved field days in which growers and technicians were taken to IPM tomato fields in eastern Guatemala and snowpea locations in the central highlands. These field days continue to the present.

Technology transfer is the first step in the broad-based sustainable NTAE production. The second step is to implement industry-wide uniform production and post-harvest handling programs consistent with U.S. Good Agricultural Practices and Food Safety guidelines for pre-inspection certification. For example, with Guatemalan snowpeas shipments to the United States, there have been a large number of port-of-entry detentions in the past. Pre-inspection protocols developed by IPM CRSP are designed to ensure the phyto-sanitary quality of snowpeas. Efforts to institutionalize a formal pre-inspection program were undertaken by AGEXPRONT, with regulation and certification of snowpeas now conducted by the Program for the Integrated Protection of Agriculture and the Environment (PIPAA), a joint AGEXPRONT–MAGA organization (Sandoval et al., 2001a,b).

The snowpea pre-inspection program (SPP) is based on ICM principles. The pre-inspection protocol includes integration of non-chemical approaches to pest management, including pest scouting and monitoring, cultural practices, physical control, and the build-up of natural controls (predators and parasitoids). To allow for adequate traceability and field certification, record keeping constitutes an important element, where pest scouting, management, fertilization, and other practices are recorded. Participating growers are visited twice a month by trained inspectors who evaluate adherence to pre-inspection guidelines. To ensure the sanitary conditions of the export pods, two snowpea samples are taken from each field, one pre-harvest (leaf samples) and one at harvest. Supervisors test samples for clorothalonil and metamidophos residues. At harvest, field supervisors issue a field certificate of clearance, which allows the product to enter packing-plant facilities. A field code is assigned to identify all product originating from specific fields. At the packing plant, plant inspectors randomly sample product, with the purpose of establishing sanitary and phyto-sanitary tolerance levels in exports. Once the product has complied with all requirements at the field and packing-plant level, a phyto-sanitary certificate, approved by the plant protection and quarantine office of the Ministry of Agriculture, is issued.

For NTAE crops more broadly, the FDA's Hazard Analysis and Critical Control Point (HACCP) and the APHIS Certified Pre-inspection Program (CPP) represent science-based risk-management approaches to safe food production. These programs are the centerpiece of successful NTAE expansion initiatives. HACCP programs are site-specific plans where producers identify 'critical control points' in food production and marketing systems, and then put appropriate monitoring and control measures in place. The GOG and the Ministry of Agriculture and USDA/FAS are now working to establish regional NTAE crop distribution centers to serve as gathering points for produce that will be properly handled and shipped to commercial outlets. These centers can further serve to help organize farmers, provide assistance in use of accepted IPM production practices, be training centers for farmer schools and workshops, and have research demonstration areas showing the latest production technology. All these activities will help focus farmers' attention to the necessity of following accepted and research-proven pest-management tactics to achieve the economic and environmental objectives and to ensure sustainability of the Central American farm community. The first center (FRUTAGRU – Associacion de Fruiticultores Agrupados) opened in November of 2002 in San Cristobal, Totonicapan, and has been successful.

Socio-economic Impacts of IPM Research and Implementation

Another important aspect of IPM programs in Central America involves evaluation of socioeconomic impacts of IPM and NTAE crop-production strategies on small-farm households in Guatemala. The objectives of this research have been to assess economic and social benefits of IPM for all members of farm families and to measure economic and social impacts of and constraints to IPM adoption by gender. Research has involved socio-economic impact assessment surveys and has examined constraints to IPM adoption and the effects of IPM adoption on the well-being of women and children in farm families. It has contributed to identifying organizational constraints to adoption and effective practice of IPM.

Results have demonstrated that most farmers are willing to adopt some, but not all, of the recommended IPM practices, and that adoption of the most difficult practices will require sustained producer contact with trained IPM technicians. In addition, incentives for producers to seek out new knowledge required to practice IPM appears to be related to availability of organizations' input supply and credit supply channels from government, coops, exporters, and banks.

Household surveys during 1998-2001 found that farmers had: (1) a high level of self and family involvement in NTAE labor, (2) a continuing need for production credit, (3) a high-level adoption of technology-based IPM but low-level adoption of scouting and other labor- and information-intensive practices, and (4) received better prices if they sold nontraditional export crops through production cooperatives or had contracts with exporters than if they marketed independently. Research uncovered a disarticulation between organizations' informational bases and producer perceptions of credit and input supply channels. This information can help in organizational recruitment and communication and can provide useful knowledge of diffusion mechanisms for IPM stakeholders throughout the research, extension, and marketing chains in order to gain broader acceptance of these IPM production strategies.

Conclusions

IPM strategies, when properly implemented and precisely managed, significantly reduce the use of pesticides to control crop-pest problems, and provide more economically sustainable and ecologically balanced production systems. Adoption of IPM by farmers can have a significant and positive effect on the socio-economic status of the farm family.

IPM research in Central America has been instrumental in establishing and institutionalizing pest-management programs, and CPP and HACCP protocols. The snowpea pre-inspection program is the first fully integrated program to be approved by the government of Guatemala and certified by APHIS. These programs and protocols will assure greater access to U.S. markets and safer food supplies for U.S. consumers while ensuring more sustainable trade. Similar pre-inspection protocols are now being developed for other NTAE crops. Regional supply consolidation and distribution centers are being established. These centers will be the focal point of all pre-inspection activities. Such programs may lead all Central American countries to establish proactive NTAE policies. Proactive policies require a 'total systems approach,' including a market-driven production and post-harvest handling strategy.

IPM programs in Ecuador have demonstrated the potential to significantly reduce pest problems and pesticide use, and to raise farm incomes for plantain and potato growers. The absence of cost-effective technology-transfer mechanisms remains a concern, however, in a country devoid of a functional public extension system. Farmer field schools have proven effective in potatoes, but relatively expensive, and perhaps non-sustainable in their current form. Efforts are underway to redesign these schools and to combine them with other approaches to reach a larger number of growers, both on the coast and in the highlands.

Acknowledgments

The authors thank the many collaborators who supported the research effort. In Ecuador, INIAP Director Dr. Gustavo Enriquez provided strong institutional support and intellectual input into the project. Greg Forbes of CIP provided excellent collaboration in plant pathology, and Aziz Lagnaoui the same in entomology. Wills Flowers at Florida A&M contributed in entomology, and Colette Harris and Sally Hamilton at Virginia Tech in gender analysis. Several researchers in Ecuador made important contributions, including Ings. Patricio Gallegos, José Ochoa, Danilo Vera, Raul Quijije, Jovanny Suquillo, Fernando Echeverría, Jorge Revelo, P. Rodriguez, Carmen Triviño, Pedro Oyarzun, Hector Andrade, Flor Maria Cardenas, and Dr. Gustavo Bernal.

Many people helped coordinate and conduct field work in the highlands and Coastal Regions. Of special mention are Luis Escudero and Ivanna Carranza. In Central America, special note goes to Glenn Sullivan of Purdue for his long-standing excellent site leadership. Rich Edwards, R. Martyn, and Jim Julian at Purdue, Ron Carroll at Georgia, and Linda Asturias deserve special note. At ICTA: D. Dardon, H. Carranza, M. Morales, O. Sierra, D. Danilo, and M. Marquez were key collaborators. Dale Krisvold, J. M. Rivera, J. C. Melgar at FHIA in Honduras, M. Mercedes Doyle, E. Barrientos, M. Bustamante, and A. Hruska at Zamorano, Honduras, and L. Caniz – APHIS-IS/Guatemala were special contributors.

References

FAO. 2003. FAOSTAT Agricultural Data. <http://apps.fao.org/page/collections?subset=agriculture>

Goldin, L., and L. Asturias. 2001. Perceptions of the economy in the context of non-traditional agricultural exports in the central highlands of Guatemala. *Culture and Agriculture* 23(1):19-31.

Hamilton, S., L. Asturias de Barrios, and B. Trevalán. 2001. Gender and commercial agriculture in Ecuador and Guatemala. *Culture and Agriculture* 23(3):1-12.

Hamilton, S., and E.F. Fischer. 2003. Nontraditional agricultural exports in highland Guatemala: Understandings of risk and perceptions of change. *Latin American Research Review* 38(3):82-110.

Julian, J.W., G.E. Sánchez, and G.H. Sullivan. 2000a. An assessment of the value and importance of quality assurance policies and procedures to the Guatemalan snowpea trade. *J. of International Food and Agribusiness Marketing* 11(4):51-71.

Julian, J.W., G.H. Sullivan, and G.E. Sánchez. 2000b. Future market development issues impacting Central America's nontraditional agricultural export sector:

Guatemala case study. *American Journal of Agricultural Economics* 82(5):1177-1183.
Julian, J., G.H. Sullivan, and G.E. Sánchez. 2001. The role of pre-inspection programs in achieving sustainability in non-traditional agricultural export markets. Integrated Pest Management in NTAE Crops, Seminar IV Proceedings, p. 33.
McDowell, H., and S. Martinez. 1994. Environmental and sanitary and phytosanitary issues for western hemisphere agriculture. *Western Hemisphere*, WR(94)2, June, pp. 73-80.
Quizon, J., G. Feder, and R. Mugai. Fiscal sustainability of agricultural extension: The case of the Farmer Field School approach." *Journal of International Agricultural and Extension Education* 8 (Spring 2001):13-24.
Sandoval, J.L., G.E. Sánchez, G.H. Sullivan, and S.C. Weller. 2001a. Transfer of IPM CRSP generated technology. Integrated Pest Management in NTAE Crops, Seminar IV Proceedings, p. 31.
Sandoval, J.L., G.E. Sánchez, G.H. Sullivan, S.C. Weller, and C.R. Edwards. 2001b. Pre-inspection model for the production and export of snowpeas from Guatemala. Integrated Pest Management in NTAE Crops, Seminar IV Proceedings, pp. 34-35.
Sullivan, G.H., G.E. Sánchez, S.C. Weller, and C.R. Edwards. 1999. Sustainable development in Central America's non-traditional export crops sector through adoption of integrated pest management practices: Guatemalan case study. *Sustainable Development International* 1:123-126.
Sullivan, G.H., G.E. Sánchez, S.C. Weller, C.R. Edwards, and P.P. Lamport. 2000. Integrated crop management strategies in snowpea: A model for achieving sustainable NTAE production in Central America. *Sustainable Development International* 3:107-110.
Weller, S.C., G.E. Sanchez, C.R. Edwards, and G.H. Sullivan. 2002. IPM CRSP success in NTAE crops leads to sustainable trade for developing countries. *Sustainable Development International* 5:135-138.
World Bank. 2001. World Development Report. Washington, D.C.: International Bank for Reconstruction and Development.

— 6 —
Developing IPM Packages in the Caribbean

Janet Lawrence, Sue Tolin, Clive Edwards, Shelby Fleischer,
D. Michael Jackson, Dionne Clarke-Harris, Sharon McDonald,
Kathy Dalip, and Philip Chung

Introduction

The Caribbean Community and Common Market (CARICOM) is comprised of fifteen member countries and territories (Antigua & Barbuda, Bahamas, Barbados, Belize, Dominica, Guyana, Grenada, Haiti, Jamaica, Montserrat, St Kitts & Nevis, St Lucia, St Vincent & the Grenadines, Suriname, and Trinidad & Tobago) and five associate members (Anguilla, Bermuda, British Virgin Islands, Cayman Islands, and The Turks & Caicos) (Figure 6-1). Other important non-CARICOM countries considered to be part of the Caribbean region are Aruba, Cuba, Dominican Republic, Haiti, Guadeloupe, Martinique, Netherlands Antilles, St. Martin, Puerto Rico, St. Maarten, and U. S. Virgin Islands. The region stretches approximately from 60° to 80° longitude and 10° to 20° latitude, spanning an arc of nearly 3000 kilometers. The region has a combined landmass of only 488 thousand km^2, three-fourths of which is Guyana and Surinam on the South American continent. In 1996, the total population was estimated at 21.5 million, 72% being located in Dominican Republic and Haiti.

The economies of the Caribbean countries are supported primarily by agriculture and tourism, and to a lesser extent by mining and manufacturing. Traditionally, the agricultural sector has contributed as much as 30% of the Gross Domestic Product (GDP) of many Caribbean countries. However, the contribution of agriculture may be even greater, since the sector is linked primarily to other business sectors such as tourism and manufacturing. In recent years, contributions of the agricultural sector to economic growth have been somewhat stagnant and in some regions even declining

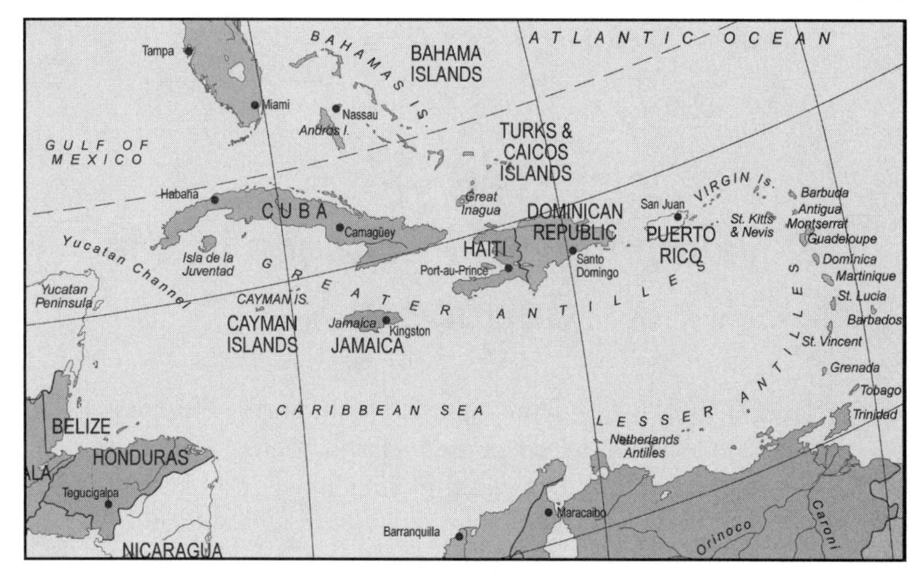

Figure 6-1. Map of the Caribbean region.

trends have been observed. The lack of effective, cost-efficient technologies for agricultural production and weak institutional support services have been recognized as contributing factors. To reverse these trends and build a more internationally competitive agricultural sector, a regional transformation program was launched. Critical elements of this program are the diversification of the agricultural sector and the development of efficient technologies for the production and marketing of new goods and services. Integrated Pest Management (IPM) is considered to be a key component for achieving this goal.

The concept of IPM was not new for the Caribbean Region, but early IPM programs focused on plantation crops such as sugar cane, bananas, coffee, and coconuts. With the recent drive to diversify the agricultural sector, many traditional garden crops (especially vegetables) grown by limited-resource farmers were identified as having potential for export. However, because of a previous lack of economic importance of these crops, information on their production was sparse, and cost-efficient technologies needed to produce adequate volumes of high quality products for export markets were largely unavailable.

The inclusion of the Caribbean Region as a participating host for the USAID IPM CRSP was a timely intervention, because research to develop programs for management of pests of horticultural and winter vegetable crops had the potential to increase non-traditional export markets from Jamaica to the United States. The Caribbean Agricultural Research and Development Institute (CARDI), a regional institute for the development and advancement of agricultural technologies across the region, was logically identified as the host institution for the IPM CRSP in the Caribbean. Since Jamaica has responsibility within CARDI for IPM, CARDI-Jamaica became the host country institution to conduct and coordinate programs of the Caribbean site of the IPM CRSP. A long-term goal from the outset was the transfer and regionalization of IPM technologies developed in Jamaica to other CARICOM countries.

Identifying and Prioritizing IPM Problems and Systems

The goals, objectives and research programs of the Caribbean site were determined through a series of on-site meetings and workshops with IPM CRSP team members and potential stakeholders. Potential stakeholders included other scientists, extension personnel, producers, processors, and distributors of crops from Government and private institutions: the Jamaican Pesticide Control Authority (PCA), the Agricultural Marketing Cooperation (AMC), the Caribbean Food and Nutrition Institute (CFNI), and two USAID Mission agricultural projects, the Agricultural Export Support Program (AESP) and the Hillside Agricultural Project (HAP). Initial meetings provided the international scientists with an overview of Jamaican/ Caribbean agriculture including crops, pests, production practices, technology transfer and marketing, and policies. Field training in Participatory Appraisal (PA) methods was organized in vegetable-producing communities in the south and north of Jamaica. These PAs provided baseline data on potential crops and communities on which to target the IPM CRSP research. Data that were collected included technical information and social/ gender and economic data on the farm households and community dynamics.

The selection of crops on which to focus research was based on detailed reviews and discussions of their export importance, the levels of pesticide residues on marketable crops, the potential for IPM development, and crop type (horticultural/vegetable). Three crops were selected: a leafy green vegetable amaranth known locally as callaloo (*Amaranthus viridis* L.), sweet potato (*Ipomoea batatas* (L.) Lam.), and hot pepper (*Capsicum chinensis*) (Figure 6-2). Historically, these crops had been grown predomi-

*Figure 6-2. Callaloo (*Amaranthus viridis*) plants from which green leaves are harvested as a leafy green vegetable.*

nantly on small holdings of fewer than two hectares, and primarily for household consumption and sale in local markets in Jamaica and other Caribbean islands. In recent years, however, their popularity in both ethnic and mainstream markets (particularly hot pepper) in the United States, Canada, and the United Kingdom, resulted in an increase in their economic importance and elevation in status to that of non-traditional export crops. The combined value of production at the farmgate for these three research crops increased by 286% from 1991–1995, an increase in overall market value from US$8M to US$30.9M (Reid and Graham, 1997).

Farming communities for IPM research were selected using several criteria, among which were production acreage, social acceptance of farmers within the production area to the use of IPM technology, and the availability of community facilitators. Two initial target communities were selected. In Annotto Bay (St Mary Parish), located in the north of the island, there was a strong interest in the cultivation of hot peppers and an established farmer's cooperative. Bushy Park and Ebony Park, both located on the southern plains of the island in Clarendon Parish, had a tradition of growing callaloo and sweet potato for local and export markets.

Baseline Data

Surveys were conducted over several growing seasons to determine farmers' perceptions of the relative ranking of pests and diseases affecting crop yield and quality, pest-management methods used by farmers, and any other constraints affecting production. The survey results, combined with crop-pest monitoring that was undertaken for two years, helped to focus research questions and were also useful for monitoring areas outside the target local communities to identify emerging factors (e.g., pests) that could be limiting production of the target crops. Interactive workshops with farmer groups, as well as interviews with individual farmers and other household members, were organized and held for each target crop. In such sessions, questionnaires and demonstrations were used to evaluate the knowledge levels of farmers about IPM and to determine critical factors affecting farmer decision-making.

Pest Composition in Target Crops

In callaloo, a range of lepidopteran species was recognized to be the most destructive pests. The species were identified as members of the families Noctuidae [fall armyworm, *Spodoptera frugiperda* (J. E. Smith.); beet armyworm, *S. exigua* (Hb); southern armyworm, *S. eridania* (Cramer)] and Pyralidae [southern beet armyworm, *Herpetogramma bipunctalis* (Fabr.); Hawaiian beet armyworm, *Spoladea recurvalis* (F.)]. Farmers also showed considerable concern for damage caused by leaf beetles (*Disonycha* spp., *Diabrotica balteata*) and mites (*Tetranychus* spp.).

Surveys of farms growing Scotch Bonnet and/or West Indian Red hot pepper varieties confirmed that yields were limited mostly by the potyviruses tobacco etch and/or potato Y. Aphids, species of which are known to be vectors of these viruses, were the most abundant arthropod pests observed. More than twenty species of aphids were recorded, with the most abundant being *Uroleucon ambrosiae*, *Aphis gossypii*, and *A. amaranthi* (McDonald et al., 2003). The broad mite (*Polyphagotarsonemus latus*) was also a major economic impediment to pepper production. After seven seasons, two gall midges (*Contarinia lycoperscisi* and *Prodiplosis longifila*) (or possibly an undescribed species) were identified as newly emergent pests. These dipteran midges not only reduce the quality of the marketable product, but also are of quarantine significance for importation into the United States. *P. longifila*, which attacks the pedicel of green mature pepper fruits and makes the fruit very susceptible to fungal attack, is considered to be the more important economically of the two species, since it is more likely to be found in hot pepper shipments. *Contarinia lycoperscisi*, on the other hand, appears limited to the flowers of the hot pepper.

The sweet potato weevil, *Cylas formicarius elegantulus* Fabricius (Coleoptera: Cuculionidae) was identified as the pest most limiting to economic production of sweet potatoes. Production losses in yields as high as 50% were reported by many growers. After several seasons of working with farmers within local target areas, a second economically important pest, identified as the sweet potato leaf beetle (*Typophorus nigritus viridicyaneus* (Crotch)), also significantly lowered quality and quantity of yields (Jackson et al., 2004). The WDS soil insect complex, consisting of wireworm, *Diabrotica*, and *Systena*, is also a significant pest in sweet potato production.

Pest-Management Practices

Farmers reported using some cultural and mechanical practices to suppress pest populations, but pesticides were by far the most commonly used practice. Pesticides were often applied weekly throughout the year, particularly to callaloo. Use of mixtures of pesticide "cocktails" was also reported. Farmers repeatedly reported failure of commonly-used pesticides to suppress pests effectively, suggesting anecdotally that a build-up of resistance to pesticides had occurred. Although many farmers reported they did not keep detailed records of production inputs, it was estimated that pesticides accounted for approximately 40% of the estimated cost of production for callaloo, 20% for sweet potato, and 12-15% for hot pepper.

For hot pepper, more than 78% of farmers used pesticides, including more than 25 different types. The most common insecticides were diafenthurion and profenofos to control the broad mite. For callaloo, all growers relied on pesticides for control of the lepidopteran larvae; thirteen different insecticides, seven fungicides, and three herbicides were recorded as being used to control various pests of callaloo. Observations and anecdotal information suggested the occurrence of resistance to pyrethroids in *Spodoptera exigua* and possibly other species. Of all sweet potato farmers surveyed, 13% used chemical pesticides in the production of their crops; seven insecticides, two fungicides, and three herbicides were recorded. However, many farmers relied on a range of alternative methods such as wood ash and cultural practices (field sanitation) for weevil management.

Surveys of callaloo and hot pepper crops on sale in local markets showed that seven of twenty-one callaloo samples (33%) and six of seventeen hot pepper samples (35%) contained unacceptable residue levels of diazinon (0.48–2.2 ppm), malathion (0.62–0.90 ppm), and prophenophos (0.94–1.10 ppm).

Designing and Testing IPM Tactics

Using accumulated knowledge, strategies were designed to develop and test appropriate IPM systems for management of pest problems identified in baseline surveys. Research objectives were identified and organized under four main components — IPM Systems Development, Pesticide Use, Residues and Resistance, Socio-economic Policy and Production Systems, and Research Enhancement through Participatory Activities. These components were linked to technology and information transfer, thereby ensuring that research needs were being targeted, and that IPM strategies were adjusted and refined to meet farmer requirements.

Identification and development of IPM strategies, as well as approaches for their implementation, varied for each crop and depended to a large extent on the information recorded during baseline surveys. Such information included identity of key pests and the taxonomic, biological, and ecological data associated with each, as well as production, marketing practices and trends, trade issues, policy issues, and regulatory considerations for each crop species.

IPM Tactics for Callaloo (Vegetable Amaranth)

Vegetable amaranth (*Amarantus viridis* L., and *A. dubius*) is grown predominantly on small holdings of less than two hectares in diversified peri-urban farms. In Jamaica, callaloo is treated as a ratoon crop, with young leaves from producing plants harvested on a weekly basis over an extended time period. Fields are established by transplanting seedlings usually grown by individual farmers from seeds produced by a "seed tree", a mature plant selected for horticultural and culinary qualities and allowed to produce seeds. The crop is thus continuously present in fields and transplant nurseries in various stages of vegetative and reproductive growth.

The numerous lepidopteran pests of callaloo were found to limit yields and to lead farmers to excessive pesticide use, often resulting in development of pesticide resistance. Hence, the team's initial focus was to develop multiple strategies that would decrease pesticide use. Early research was done to identify sampling methods for field-pest scouting, to develop action thresholds, and to devise early warning devices for monitoring pest populations. One of the first steps was to clarify the taxonomy of the lepidopteran complex. Identification manuals were developed that could be used in training programs (Clarke-Harris et al., 1998). The introduction of pesticides with newer selective chemistries was considered a priority in order to replace many of the persistent, broad-spectrum pesticides commonly used by growers. For long-term strategies, various non-chemical practices such as the use of exclusion cages and biological agents were identified as promising tactics to be evaluated in the second phase of the development of IPM technologies. A knowledge base was established regarding pesticide resistance and resistance management. The advances in IPM tactics in controlling pests in this crop were recognized as a model to advance IPM in crucifers and leafy greens that are subjected to high pesticide inputs in both the eastern and western parts of the Caribbean.

IPM Tactics for Hot Pepper

The main research focus on hot pepper was to develop IPM technologies to manage the virus complex, the limiting factor in the production of Scotch Bonnet pepper. Early investigations sought to determine basic information on the range, severity, and identity of the viruses and their associated vectors (composition, population biology) as well as the epidemiology of the viruses (McDonald, 2001; McDonald et al., 2003). These data were then used to develop a more ecologically based strategy to reduce viral transmission (e.g., use of exclusion in seed beds, stylet oils, roguing) (McDonald et al., 2004a, b). However the impact of this approach was viewed as being long-term with relatively few short-term gains for growers. Hence, studies were also conducted to identify and introduce varieties of hot pepper tolerant to the viruses. The West Indian Red hot pepper variety was adopted by farmers as a stop-gap approach to minimize the crippling effects the viruses were having on production of the more desirable Scotch Bonnet pepper.

The emergence of gall midges as serious pests of hot pepper also presented new challenges for basic research on pest biology and ecology. Initial studies evaluated a combination of newer pesticide chemistries, combined with cultural practices (such as fruit-stripping), with the long term goal of developing more biologically based IPM strategies.

In consideration of the long-term goal of transferring IPM CRSP technologies to other countries within the Caribbean, research was initiated by the USDA to develop hot pepper varieties with resistance to rootknot nematodes for countries such as St Kitts where this pest was a limiting factor in the production of hot peppers and other vegetable crops. The aim of the breeding program was the transfer of southern root-knot nematode resistance from Scotch Bonnet pepper into Habanero pepper and West Indian Red hot pepper. In addition, the potential of the resistance to withstand the high temperatures in tropical countries and endemic races of rootknot nematodes was assessed in field trials in St Kitts.

IPM Tactics for Sweet Potato

Sweet potato provided an opportunity for immediate introduction of a component of a biologically based IPM strategy because it was a crop for which, in comparison to callaloo and hot pepper, there were sufficient baseline data on the key pests. The fact that data were available from research conducted previously by members of the IPM-CRSP team from Jamaica and the USA, as well as worldwide (Palaniswamy et al., 1992; Talekar, 1998) allowed immediate evaluation of potential IPM tactics, which included cultural practices, growing resistant varieties propagated by vegetative cuttings, and the use of pest behavioral-modifying chemicals (pheromones).

To varying extents, the investigations were conducted on farms with the farmers' involvement. This participatory approach enabled the team to receive feedback on the acceptance by farmers of the technology proposed. Once validation of the IPM packages occurred, information was disseminated to extension personnel and farmers within and outside the local communities.

Callaloo Case Study

Callaloo posed a particularly interesting challenge for IPM development for several reasons apart from those mentioned above (high pest pressure and indiscriminate use of pesticides). First, very little information on the taxonomy, biology, or ecology of the key pests affecting callaloo existed. Taxonomic studies revealed that five lepidopteran species were part of the pest complex, and enabled the development of an identification guide for training growers (Clarke-Harris et al., 1998). Any management plan would need to work for all five species. Second, at the time of the inception of the project the crop was on the pre-clearance list of the USDA-APHIS, with few problems of pest interceptions (i.e., pest residues "hitchhikers" on marketable produce at U.S. ports of entry). However, subsequently high levels of pest interceptions at U.S. ports caused the crop to be removed from the pre-clearance list, which had implications for the quality of the marketable product on arrival at U.S. ports. These factors intensified the need for developing not only pre-harvest management strategies but also efficient post-harvest technologies.

To guide farmers rapidly in improved methods and kinds of pesticide application, priority was given to the development of an action threshold for the five species of Lepidoptera affecting callaloo. In the absence of empirical data, the collective experience of the IPM team was used to identify a provisional threshold of 2.5 larvae per plant. In replicated field studies, the use of this threshold resulted in an eight-fold reduction in pesticide applications compared to the grower standard of regular weekly sprays. Using these data, various action thresholds were developed and tested. The data were used to refine pesticide application timing guides.

After these initial studies, a sampling plan was developed to assist farmers in methods of scouting fields and improving decision-making skills with regard to pesticide applications. Initial studies for the development of the sampling plan included determining the within-plant distribution of each pest species, and monitoring populations of the five key lepidopteran species. From the analysis of these data, a six-leaf sample unit consisting of leaves from the inner and outer whorl of the callaloo plant was selected as a sample to assess the incidence of all caterpillar species. Sequential sampling plans were generated from a negative binomial frequency distribution model. Inputs of an action threshold of one larva per plant (per 6 leaf sample) and varying levels of error rates for the decision-making process were assessed, using subjective knowledge of farmer tolerances and potential sampling labor inputs. A sampling plan using a minimum of 10 samples, to reduce decision error, was compared to a fixed plan using a maximum of 25 samples, based on the time required (45 minutes to 1 hour). The use of the minimum sampling plan was validated on 32 farms. The same management decision was reached on 87.5% of the farms with the sampling plan as with the fixed plan of sampling 25 plants. The sampling plan recommended taking additional samples for 9.4% of the sites, and gave inaccurate decisions at 3.1% of the sites, while reducing the number of samples by 46%. Insecticide frequency of use was reduced by 33% to 60% when management decisions were based on sampling data compared to previous grower-standards, with no increase in crop damage.

The commonly used pyrethroid insecticides gave only low levels of pest suppression, and crop damage remained high or variable (10-46%). We thus evaluated biorational pesticides, among which were ecdysone agonists (e.g., tebufenozide, Dhadialla et al., 1998), the semi-synthetic microbial metabolite emamectin benzoate from *Streptomyces avermitilis* (Lasota and

Dybas, 1991), spinosyns, which are microbial metabolites from *Saccharopolyspora spinosa* (DowElanco, 1997), and insecticides with growth regulator activity produced from neem (National Research Council, 1992). Efficacy trials with these novel modes-of-action pesticides demonstrated dramatically improved lepidopteran control compared to the traditionally used insecticide lambda cyhalothrin. In field trials, the biorational insecticides spinosad, tebufenozide, and emmamectin benzoate gave the lowest levels of damage and crop loss when compared with standard insecticides used by growers.

Clearly, the introduction of biorational insecticides would improve pest-control efficacy and farm-worker safety. Much of this crop is produced in a peri-urban agricultural setting, with direct human consumption of fresh leafy material. Thus introduction of the new, selective pesticides for commercial use would also improve food safety. However, development of pesticide resistance in pest populations could be predicted if these materials were used indiscriminately. An investment in training in methods for measuring and managing pesticide resistance occurred concurrently with the field work, and the development of a sampling plan provided a mechanism of limiting spray frequencies dramatically, thus helping to manage resistance and farm-worker safety issues. Moreover, transitioning to biorational insecticides opens the potential for integration of conservation biological control.

Ideally, minimizing or totally eliminating pesticide use in the production of callaloo is a long-term goal. Therefore, the use of alternative practices such as exclusion barriers was evaluated to determine their potential to exclude pests from plants. The potential of new horticultural systems involving methods of pest exclusion opens opportunities for new and profitable small businesses. Although the land area of callaloo grown under cover would probably be a relatively small percentage of the total, there appears to be potential to eliminate insecticides and market the produce directly.

The interception of pest residues on callaloo leaves and stalks continued to be a major impediment to export. Post-harvest technology for callaloo at packing houses is not standardized in regard to the solutions commonly used to wash off pests, and the method of manual agitation of stalks in the solution to disinfest them is somewhat inefficient. Designing and developing a washer basket and optimal concentrations of washing solutions assisted in reducing the incidence of insect remains on the marketable product. Such protocols have been developed for use in processing plants.

Hot Pepper Case Study

In Jamaica, the main viruses affecting pepper are aphid-borne and non-persistently transmitted. *Tobacco etch virus* (TEV) and to a lesser extent, *potato virus Y* (PVY) both of the family Potyviridae are the main mosaic-causing viruses found on hot-pepper farms surveyed in Jamaica (McGlashan et al., 1993; Myers, 1996). Only a few aphids are required to spread these viruses in the field. Consequently, one cannot develop economic thresholds for the vectors. Furthermore, the earlier the crop stages at the time of infection, the greater the loss in yield. It is, therefore, imperative that virus disease management begins before the crop is planted. Pepper nurseries should be protected with aphid exclusion covers, and/or be established away from possible sources of inoculum, such as mature pepper fields.

Field spread of TEV is closely associated with aphid flights (transient aphids). Aphid flight in St. Catherine is seasonal; the greatest abundance and diversity of aphid species occurs from mid-September through mid-May. Five known TEV vectors were found on pepper farms in St. Catherine: *A. gossypii*, *A. craccivora*, *A. spiraecola, L. erysimi,* and *M. persicae*. Flight of two aphid species with unknown TEV vector status, *A. amaranthi* and *U. ambrosiae* complex, were also closely associated with increase in TEV incidence in one field study (McDonald et al., 2003). Pepper when grown for the winter market is transplanted during August and September; hence, the crop is most vulnerable to virus infections when aphid flights are high. Farmers should either plant their crops earlier or implement strict management practices to delay and/or reduce TEV into the crop.

Pepper is often grown in overlapping crop cycles in Jamaica, and old pepper fields are often abandoned but not destroyed. These practices provide a large reservoir of virus inoculua for infection of newly established pepper fields. Weeds are a severe problem in most pepper fields, especially during periods of heavy rainfall. Weeds compete with crops for nutrients and light, and can harbor viruses as well as aphid vectors. Proper field sanitation alone could greatly delay the introduction, and reduce the incidence, of viruses in new susceptible pepper fields.

Some weeds and crops within and around fields are hosts of common aphid species. Colonies of *M. persicae* were found on peppers grown close to *Brassica* sp.; *A. gossypii* colonizes pepper as well as several vegetable crops, fruit trees, and weeds in pepper fields; *U. ambrosiae* forms dense colonies on

Parthenium hysterophorus; *A. amaranthi* colonizes *Amaranthus* spp.; and *T. nigriabdominalis* colonizes the roots of grasses. *Amaranthus* spp., *P. hysterophorus*, and grasses are major components of the weed community on pepper farms. Many farmers in St. Catherine grow a species of amaranth as a vegetable crop adjacent to pepper fields.

The dynamics of the infection of Scotch Bonnet pepper with TEV was studied in field plots by McDonald (2001). Primary infections were random throughout the plot and originated from outside pepper fields. Thereafter, field distribution of TEV became spatially correlated and occurred mainly by secondary spread within the field. TEV-infected plants increased logistically within pepper plots; therefore the rate of spread is greatest during the middle of the infection cycle. Plots less than one hectare were totally infected within ten weeks.

This knowledge suggests that roguing of infected pepper plants might be an effective measure of control only in the first two to three weeks of infection, a time at which only well-trained eyes can detect early symptoms of virus. Roguing would also require that the nearest neighbors of infected plants be removed even if they have no symptoms because they too could be infected (Broadbent, 1969). Roguing requires good sanitation practices; rogued plants should be removed from within the plot and be promptly destroyed so that they do not become a source for the virus or any vectors they might be harboring (Broadbent, 1969). Replacing rogued plants with healthy ones would increase overall productivity per acre, but the new plants are often less hardened and more susceptible to virus infections than other plants in the stand. Farmers might not have the time required to conduct efficient roguing.

Increasing the field size or the planting density might decrease the rate of spread of TEV. There would be more plants per unit area, requiring more time to get infected. Additionally, increased field sizes would reduce the edge effects from the tendency of immigrant aphids to land on the borders of fields. Scotch Bonnet peppers grown at a density of 0.6 x 0.6 m produced more fruit than those grown at a density of 0.9 x 0.9 m (McGlashan, personal communication). Intercropping and use of barrier crops could also help to delay the spread of TEV in pepper fields. It would be advisable to plant barrier crops so they can get established before peppers are transplanted. Both barrier crops and intercrops need to be non-hosts to the virus and its vectors.

Scotch Bonnet pepper inoculated with TEV during the vegetative stage and at the onset of flowering (during the first month after transplanting) exhibited severe retardation in growth and yield although the developmental stages were not inhibited (McDonald et al., 2004a). Plants inoculated 7 and 28 days after transplanting were shorter, covered less ground area, and produced fewer and smaller fruit than did the control (Figure 6-3). Scotch Bonnet pepper plants inoculated with TEV after the first stage of flowering (approximately two months after transplanting) were not significantly different from uninoculated plants in size or in yield. Increases in Scotch Bonnet pepper yield, therefore, can be obtained by protecting plants from TEV infection during the seedling stage through first stage of flowering.

Covering nursery beds or seedling crates with aphid-exclusion cages made with ordinary sheer curtain material is an economical way for farmers to protect pepper seedlings from aphid-borne viruses before transplanting. Reflective mulch made from aluminum-coated plastic and JMS Stylet-Oil® together have been proven to be more effective in reducing and delaying the spread of TEV and other viruses than stylet oil alone (Mansour, 1997). Stylet-Oil® alone, applied once weekly with a backpack mist blower (low volume, single nozzle, and pressures of about 1000 kPa), delayed field spread of TEV by seven days and reduced TEV incidence by 24% in sprayed plots when compared to unsprayed plots (McDonald, 2004b). Stylet-Oil® and reflective mulch together delayed the incidence of TEV in pepper plots for more than two months, even with high inoculum pressures of 33-67% infection from surrounding plants.

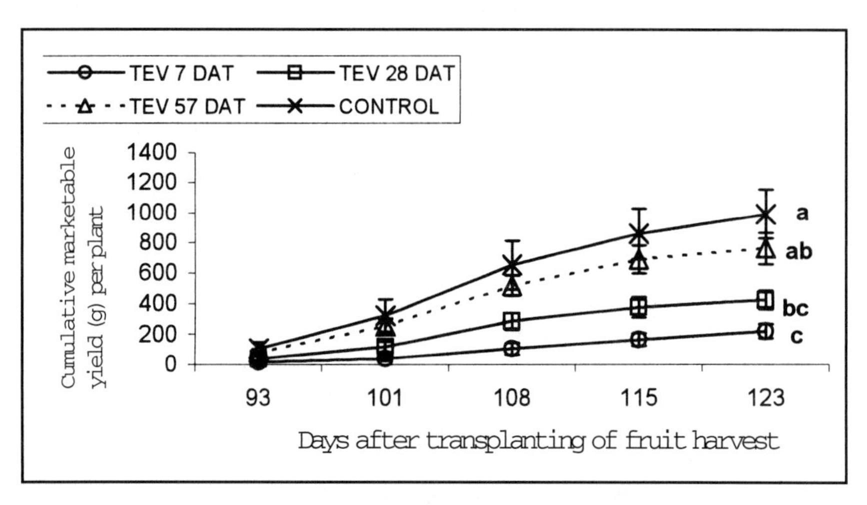

Figure 6-3. Yield of Scotch Bonnet pepper fruit on plants inoculated with tobacco etch virus (TEV) at 7, 28, or 57 days after transplanting.

Plastic mulches will require the use of drip irrigation. Straw has been shown to repel whiteflies (Nitzany et al., 1964). Myers (1996) reported that straw mulch can delay the incidence of TEV in pepper in Jamaica. Straw is biodegradable and less costly than plastic mulch, and could be substituted if proven to work well in suppressing the spread of TEV in pepper. However, the reflective properties of straw are not as great; thus, straw may not repel aphids as well as plastic mulches do.

Simons et al. (1995) recommend that Stylet-Oil® applications begin after five to six winged aphids are caught in 9-inch water traps over 24 hours and that frequency of applications be increased to twice per week when at least 15 aphids are caught in 24 hours, and after heavy rains. Given the small size of the typical Jamaican pepper farm, and the typically low numbers of aphids trapped in pepper fields in Jamaica, it would be advisable that farmers start spraying Stylet-Oil® as soon as the seedlings are transplanted.

Sweet Potato Case Study

Successful sweet potato IPM technologies have been described from Asia (Talekar, 1988, 1991; Jansson and Raman, 1991; Palaniswamy et al., 1992; Hwang, 2000), Africa (Smit and Odongo, 1997), Cuba (Alcázar et al., 1997; Morales-Tejon et al., 1998; Lagnaoui et al., 2000; Maza et al., 2000), Haiti, and the Dominican Republic (Alvarez et al., 1996). The low-input IPM technology package extensively used in Asia (Telekar, 1988, 1991) was selected as being the most readily adaptable for our purposes in Jamaica, and suitable for reducing damage caused by the key sweet potato pest, the sweet potato weevil. The technologies included the use of selected cultural practices (field sanitation — removal of crop residues and alternate hosts, quick harvesting, and use of clean planting material) (Jackson et al., 2002), and mass-trapping of weevils using a sweetpotato weevil sex pheromone (100mg) (Heath et al., 1988) (Figure 6-4).

Following slight adaptations to meet local needs, the IPM technology was validated on seven farms in the target area of Ebony Park. Validation of the IPM technology included three phases: (i) gathering of baseline data on economic damage due to the weevil and production practices employed by farmers, (ii) introduction of an IPM technology package to farmers using a modified farmer field school approach, and (iii) impact assessment of the introduced technology after three seasons.

The impact of the technology package was evaluated on farms where farmers had been trained in the IPM practices and had utilized them for three seasons (n=7), and on equivalent neighboring farms where the farmers did not practice IPM (n=11). Factors that were assessed included: assessing weevil populations, cultural practices employed, trap maintenance (where applicable), and resultant crop losses. Significant differences in yield losses caused by the weevil were recorded between farmers who utilized the IPM approach and those who relied only on cultural practices ($p<0.007$), with damage from the two approaches averaging 4% and 13%, respectively (SED=3.00/0). Similar trends were observed in pheromone trap captures of male weevils. Mean weevil catch per 0.1 ha of sweet potato was 779 (SE± 139) for non IPM farms and 22 (SE±149) for IPM farms, respectively.

Interestingly, no differences were observed in the numbers and types of cultural practices utilized by farmers who had been exposed previously to IPM training and those who had not been exposed. The most commonly practiced cultural control method reported by farmers was frequent irrigation to avoid soil cracking (94%), and the removal of old roots and vines from the field (83%). Many of the additional cultural practices we recommended were not adopted by farmers due to several socio-economic factors. For example, removal of vines from old fields was not practiced because farmers also had livestock and these fields were used as sources of fodder for animals. Farmers also used old fields as a source for planting material. The unavailability of tractor services also made proper field sanitation difficult. These observations, as well as those relating to the use of pheromone traps, were used to modify the recommendations and to identify topics for farmer training, including cost-benefit analyses of field sanitation, and trap maintenance in relationship to trapping efficiency.

Despite these shortcomings, the IPM CRSP model compared favorably with other IPM programs tested in the Caribbean. Both the Jamaican and Cuban IPM programs (Alcázar et al., 1997; Maza et al., 2000) reported 2-3-fold reductions in pest damage over conventional techniques. However, the Cuban model relies more on pest-resistant or tolerant varieties (short-season types and cultivars with deep root formation) and biological control agents (fungal pathogens and predators).

During the development of the IPM package for Jamaica, another important sweet potato insect pest was identified on farms where the IPM program for sweet potato weevil management was implemented. Initially, this new pest was described as a "white grub" because the damage it caused was similar to the channeling of the surface of sweet potato roots caused by other white grub species, such as *Plectris aliena* Chapin and *Phyllophaga* spp.

(Coleoptera: Scarabaeidae) (Cuthbert, 1967). However, the larvae causing this damage were not scarabs but chrysomelid larvae, later identified by Jackson et al. (2004) as the sweet potato leaf beetle (SPLB), *Typophorus nigritus viridicyaneus* (Crotch) (Coleoptera: Chrysomelidae, Eumolpinae) (Arnett et al., 2002). The adult of *Typophorus* sp. had been identified as a pest of sweet potato foliage in Jamaica in early pest analyses, but was not at that time associated with the larval damage on sweet potato roots. In later years, research on the sweet potato leaf beetle in Jamaica has been an important effort of the IPM CRSP for the Caribbean Site. Through grower surveys over the last few years, the sweet potato leaf beetle has emerged as a predominant pest species of sweet potato in certain parishes of Jamaica. On some parts of the island, over 80% damage to roots has been reported. Fipronyl and an organic garlic extract also were effectual in controlling this pest, but effectiveness increased when used in combination with the resistant sweet potato genotypes (Jackson et al., 2004).

Other IPM tactics for the management of the sweet potato pests included the development and use of multiple pest-resistant sweet potato cultivars. In collaboration with the USDA-ARS in South Carolina, efforts focused on developing and screening high-yielding, red-skinned, cream-fleshed sweet potatoes from the USDA and local Caribbean varieties in the germplasm collection of the Jamaica Ministry of Agriculture (MINAG) (Jackson et al., 1998). Many of the USDA varieties were bred for resistance to root-knot nematodes, diseases, and insects, including both the leaf beetle and the sweet potato weevil. Compared to the popular local sweet potato variety 'Sidges', several USDA-developed sweet potato clones and Caribbean varieties — White Regal (Bohac et al., 2001), the Cuban variety 'Picadito', PI 531116, TIS 24-98, and the local variety 'Fire-on-Land' — demonstrated high levels of resistance to the sweet potato leaf beetle, sweet potato weevil, and/or the wireworm-*Systena-Diabrotica* (WSD) soil insect complex. Yields were either comparable to or greater than yields from the local grower standard variety. These varieties were evaluated on pilot farms where the farmers determined their performance and culinary and taste characteristics. Preliminary results from informal culinary tests indicated that the farmers were pleased with the quality of the potatoes produced (shape, texture, color of skin and flesh).

Integrating Social and Gender Analyses

Socioeconomic Factors Influencing the Adoption of IPM

Key factors identified early in the program that were likely to limit the introduction and effectiveness of IPM technologies were the inability of the farmers to diagnose pest problems and a lack of understanding of the role of natural enemies in the suppression of pests. These observations explained, at least in part, the inappropriate selection and use of pesticides and pest-management techniques that were commonly observed.

Social, economic, policy, and institutional systems (human systems) can present overwhelming barriers to the implementation of IPM practices. Therefore, efforts were made to identify components of human systems that may limit the adoption of the IPM technologies that were developed.

A key factor identified as a constraint to the adoption of IPM was found to be related to gender equity (G. Schlosser, 1998; T. Schlosser, 1998). Studies conducted in communities associated with the IPM strategy-development experiments indicated that women are producers and pest-management decision makers for the target crops (e.g., selection and purchase of pesticides), and are thus key to the successful dissemination of IPM technologies. It was found, however, that women farmers were less likely to receive relevant technical assistance or to be involved in technology development. Women who do not farm were shown to have considerable control over farm incomes, a factor that could have an impact on the type and quantities of pesticides being purchased. Women may be more pesticide-dependent than men are since they have limited access to IPM technology and are therefore less likely to be aware of and utilize alternatives to chemical pesticides. In response to this information, concerted efforts were made to include women in IPM field activities and to disseminate gender-disaggregated information through the extension arm of MINAG, the Rural Agricultural Development Authority (RADA).

The potential impacts of various domestic and trade policies of Jamaica, as well as the trade policies of the USA, on the production practices of farmers adopting IPM CRSP production technologies were investigated. Research indicated that IPM technologies make crop production more profitable than conventional systems do, i.e., increases in yields and net returns (Ogrodowczyk, 1998). Simulation models suggested that even the lowering of trade barriers, by encouraging the pre-clearance of vegetables prior to export, eliminating water subsidies and credit, lowering the common external tariff, and increasing the exchange rate, would not reduce the profitability of IPM adoption. These data were therefore encouraging and were used in technology dissemination sessions.

Facilitating Linkages

Through workshops, meetings, and seminars, intra- and inter-regional linkages were achieved among persons within the production sector — including farmers, extension workers, and researchers from varied disciplines (pathologists, entomologists, agronomists, weed scientists), marketing agents, quarantine personnel from government and non governmental institutions, processors, and distributors. Such approaches assisted in: (i) analyzing, refining, and disseminating IPM technologies and recommendations; (ii) streamlining the research efforts of research institutions to avoid duplication; (iii) strengthening the resource base to advance technology development; (iv) creating multidisciplinary research teams; and (v) short and long term training for both Caribbean and U.S. nationals.

Technology Transfer

Meetings with RADA and MINAG personnel helped to identify communities for IPM training and broadened the scope of the IPM CRSP beyond the original primary sites. Meetings were held with farmers throughout growing seasons, not only to train and disseminate IPM technology, but also to receive feedback on the appropriateness of the recommendations made. Various methods were used to disseminate IPM technologies to farmers, all having an interactive nature and a principle-based approach to learning.

Using sweet potato as an example, a four-step approach was used to develop and disseminate sweet potato weevil IPM technology to farmers. The steps were an initial baseline survey, technology development, training sessions, and impact assessment. To introduce farmers to the technology, introductory seminars of 40–60 agricultural officers/students and farmers were held initially and were used to validate the information collected during the baseline survey and also to discuss IPM principles and the sweet potato weevil IPM technologies available to them. The scientific rationale behind the recommended practices was emphasized. Subsequent to these larger sessions, small-group gatherings were held for farmers to provide hands-on experience in the construction of pheromone traps and to identify problems experienced with the technology. More than 200 persons were trained in these technologies.

Researcher Training

Skills in IPM have been developed through short- and long-term training programs held both in the Caribbean and the United States. Short-term research training topics included pest sampling, the measurement and management of pesticide resistance, statistical analyses, acarology taxonomy, molecular biology methods, informatics — email, internet and world wide web, and GIS technology. These training sessions strengthened the skills and capabilities of the scientists in the Caribbean. The use of knowledge acquired and skills in these areas has been evident in both CRSP and non-CRSP research programs.

Eight students have received graduate training in the United States: one Jamaican national (M.S. and Ph.D. in Entomology and Virology) and six United States nationals (M.S. in Geography, Sociology and Gender, Economics). The Caribbean national now serves as a vital resource person as MINAG's Post-entry Quarantine Officer. A Jamaican national is completing a Ph.D. program in weed management at the University of the West Indies (UWI) in Jamaica.

Linkages and Training in Information Technologies in IPM

In the PROCICARIBE network in the Caribbean, the Caribbean Integrated Pest Management Network (CIPMNET) was established for IPM information exchange. It was a natural extension of the IPM CRSP to assist CIPMNET in the development of information technologies. Through IPM CRSP training programs to agricultural scientists, in-country capabilities for geographic information systems (GIS), global positioning systems (GPS), and rapid web-based mapping was developed and linked to UWI. Training was initiated in the western Caribbean (Jamaica), and regionalized with a second workshop in the eastern Caribbean (Trinidad) with the Centre for Geospatial Studies, UWI-St Augustine.

Along with training, the IPM CRSP prototyped web-based pest monitoring, mapping, and display of both maps and time-series graphics using the gall midge of hot pepper as the model system. A multi-agency workshop (RADA, MINAG, CARDI, Penn State, UWI, Agro Grace, H&L Agri & Marine Co., Agricultural Chemicals Plant, Pesticide Control Authority) reviewed and made design recommendations for application software. Pepper infestation rates were determined from six targeted parishes on a 2-week basis, based on collection and dissection of fruit by three MINAG laboratories. Spatially referenced, password-protected data input occurred via Active Server Pages forms on the web. Data were routed through the RADA server, automatically updating a spatially referenced database. A Delphi application downloaded this database and created new maps of pest-infestation rates, with clickable views of the time series at local sites. All maps and time series were ported to a public website, <radajamaica.com.jm>. This type of information infrastructure can be useful for future pest surveillance, monitoring, and management programs, including by other IPM CRSP sites, and for help in tracking invasive species, including species that influence U.S. agriculture.

IPM also influences economic development and trade. GIS was used to analyze the complex relationships among environmental, horticultural, socioeconomic, and pest-management variables for the gall midge in Jamaica (Williams, 2001).

Regionalization of IPM

A regionalization component of sweet potato IPM was initiated in 1998. It has involved variety trials, on-farm demonstrations, and workshops to transfer IPM technology developed in Jamaica for management of the sweet potato weevel with sex pheromone trapping to farmers in Antigua and St Kitts & Nevis. The response of farmers to these training workshops was very positive, and they left workshops with an increased understanding of the pests and their management. Having seen the potential impact of the trapping technologies visually, they were prepared to implement some of the cultural practices on their farms.

A two-day workshop, *Development of IPM in Leafy Vegetables that Currently Experience High-Pesticide Input*, was held in June 2002 in collaboration with the Ministry of Agriculture, Trinidad & Tobago. Nine researchers from Barbados and Trinidad & Tobago were trained in a stepwise approach towards the development of IPM strategies for a pesticide-reliant system, primarily in cabbage and crucifer production systems, and field experiments to test the strategies were established. Trinidad, under the project *Ecological Crop Management Using Farmer Participatory Approaches*, which had been developed by CIPMNET, had previously conducted a PRA of cabbage farmers. This same survey instrument was used in Barbados, Tobago, and Jamaica, and additional analyses were done in Trinidad.

Conclusions

Although Jamaica has been the primary site for the IPM-CRSP in the Caribbean region, it has had an impact throughout the Caribbean. The IPM CRSP was recognized by Frank McDonald, former CARDI director for IPM, as a catalyst for a number of IPM initiatives throughout the region. Its presence was integral to the formation of CIPMNET, a regional coordinating unit that has as its purpose "the implementation of IPM strategies that encompass cost-effective measures of prevention, observation, and intervention in a holistic approach that enhances profitability and environmental protection while maintaining pest populations at levels above those causing economically unacceptable damage or loss." CIPMNET's approach is to work among national programs with access to international organizations as needed, to develop appropriate IPM strategies for local circumstances, and to integrate IPM strategies in product marketing and sales. Information on CIPMNET is accessible through the PROCICARIBE website <www.procicaribe.org>.

Notes

The core research for the Caribbean Site was conducted by a multidisciplinary team of scientists and outreach personnel from various local, regional, and U.S. institutions. The IPM-CRSP team included entomologists, plant pathologists, virologists, nematologists, plant breeders, geneticists, weed scientists, soil toxicologists, agronomists, sociologists, economists, and production and marketing specialists. Locally, scientists were primarily from CARDI (K. Dalip, D. Clarke-Harris, J. Lawrence, R. Martin, F. McDonald, S. McDonald, J. Reid), the Jamaican Ministry of Agriculture (MINAG) (J. Goldsmith, D. McGlashan), the Jamaican Rural Agricultural Development Authority (RADA) (P. Chung), the University of the West Indies (UWI) (D. Robinson, V. Stewart), the Inter-American Institute for Cooperation on Agriculture (IICA), and USDA-APHIS. Collaborating scientists from the United States were from several institutions, including Virginia Polytechnic Institute and State University (VT) (V. Fowler, L. Grossman, S. Hamilton, W. Ravlin, A. Roberts, S. Tolin, H. Warren), Ohio State University (OSU) (C. Edwards), Pennsylvania State University (PSU) (S. Fleisher, C. Pitts), University of California at Davis (UC-D)(J. Momsen), Lincoln University (LU) (F. Eivasi), and the USDA-ARS U.S. Vegetable Laboratory, Charleston, South Carolina (J. Bohac, R. Fery, H. Harrison, D. M. Jackson, J. Theis). Management and coordination of research and administrative activities were the responsibility of the U.S.-

based Site Chair (Frieda Eivazi [LU], 1993-1995; William Ravlin [VT/OSU], 1995-1998; and Sue Tolin [VT], 1998-2004) and the CARDI-based Site Coordinator (J. Lawrence, 1993-1999; D. Clarke-Harris, 2000-2004). The IPM-CRSP website, <http://www.ag.vt.edu/ipmcrsp/annrepts/arintro.html>, contains in depth information in the annual reports for each of the years of the project, summarized in this chapter.

References

Alcázar, J., F. Cisneros, and A. Morales. 1997. Large-scale implementation of IPM for sweet potato weevil in Cuba: A collaborative effort. Pp. 185-190 in *Progr. Rept., Intl. Potato Center* (CIP). Lima, Peru, 1995-1996.

Alvarez, P., P. Escarraman, E. Gomez, A. Villar, R. Jimenez, O. Ortiz, J. Alcázar, and M. Palacios. 1996. Economic impact of managing sweet potato weevil (*Cylas formicarius*) with sex pheromones in the Dominican Republic. Pp. 83-94 in *Case Studies of the Economic Impact of CIP Related Technology*, ed. T. Walker and C. Crissman. Lima, Peru: Intl. Potato Center (CIP).

Arnett, R.H., Jr., M.C. Thomas, P.E. Skelley, and J.H. Frank. 2002. *American Beetles, Volume 2*. Boca Raton, Fla.: CRC Press.

Bohac, J.R., P.D. Dukes, Sr., J.D. Mueller, H.F. Harrison, J.K. Peterson, J.M. Schalk, D.M. Jackson, and J. Lawrence. 2001. 'White Regal', a multiple pest- and disease-resistant, cream-fleshed, sweetpotato. *HortScience* 36: 1152-1154.

Clarke-Harris, D., and S.J. Fleischer. 2003. Sequential sampling and biorational chemistries for management of lepidopteran pests of vegetable amaranth in the Caribbean. *Journal of Economic Entomology* 96: 798-804.

Clarke-Harris, D., S. Fleischer, and A. Fender. 1998. *Major pests of callaloo. Identification guide.* University Park, Pa.: Pennsylvania State University. 16 pp.

Cuthbert, F. P., Jr. 1967. *Insects affecting sweet potatoes.* USDA Agric. Handbook 329.

Dhadialla, T.S., G.R. Carlson, and D.P. Le. 1998. New insecticides with ecdysteroidal and juvenile hormone activity. *Annual Review of Entomology* 43: 545-569.

DowElanco 1997. Introducing naturalyte insect control products. *Down to Earth*, vol. 52. DowElanco, Indianapolis.

Heath, R.R., J.A. Coffelt, F.I. Proshold, P.E. Sonnet, and J.A. Tumlinson.1998. (Z)-3-Dodecen-1-ol (E)-2-Butenoate and its use in monitoring and controlling the sweetpotato weevil. United States Patent 4,732,756. March 22.

Hwang, J.S. 2000. Integrated control of sweet potato weevil, *Cylas formicarius* Fabricius, with sex pheromone and insecticide. Pp. 25-43 in *Control of Weevils in Sweet Potato Production,* ed. C. Chien-The. Pennsylvania State University. Proceed. Internat. Symp., Satellite Session, 12th Symp. Internat. Soc. Trop. Root Crops, 11-15, Tsukuba, Ibaraki, Japan.

Jackson, D.M., J. Bohac, J. Lawrence, and J.D. Muller. 1998. Multiple insect resistance in dry-fleshed sweet potato breeding lines for the USA and the Caribbean. Pp. 274-280 in *IPM CRSP Research, Proceedings of the Third IPM CRSP Symposium, 15-18th May 1998.*

Jackson, D.M., J.R. Bohac, K.M. Dalip, J. Lawrence, D. Clarke-Harris, L. McComie, J. Gore, D. McGlashan, P. Chung, S. Edwards, S. Tolin, and C. Edwards. 2002. Integrated Pest Management of Sweet potato in the Caribbean. Pp. 143-154 in *Proceedings of the First International Conference on Sweetpotato Food and Health for the Future*, ed. T. Ames. 26-30 Nov., 2001, Lima, Peru. *Acta Horticulturae* 583.

Jackson, D.M., J. Lawrence, K.M. Dalip, D. Clarke-Harris, J.R. Bohac, P. Chung, S. Tolin, and C. Edwards. 2004. Management of sweetpotato leaf beetle, *Typophorus nigritus viridicyaneus* Crotch (Coleoptera: Chrysomelidae), an emerging pest in the Caribbean. *Florida Entomologist* (In review).

Jansson R.K. and K.V. Raman. 1991. *Sweet Potato Pest Management: A Global Perspective*. Boulder, Colo.: Westview Press.

Lagnaoui, A., F. Cisnerros, J. Alcázar, and F. Morales. 2000. A sustainable pest management strategy for sweet potato weevil in Cuba: A success story. Pp. 3-13 in *Control of Weevils in Sweet Potato Production*, ed. C. Chien-The. Proceed. 12th Intl. Symp. Intl. Soc. Trop. Root Crops, 11-15 Sept., Tsukuba, Japan.

Lasota, J.A., and R A. Dybas. 1991. Avermectins, a novel class of compounds: implications for use in arthropod control. *Annual Review of Entomolology* 36: 91-117.

Mansour, A.N. 1997. Prevention of mosaic virus diseases of squash with oil sprays alone or combined with insecticide or aluminum foil mulch. *Dirasat. Agric. Sciences* 24: 146-151.

Maza, N., A. Morales, O. Ortiz, P. Winters, J. Alcázar, and G. Scott. 2000. Economic impact of IPM on the Sweet Potato Weevil (*Cylas formicarius* Fab.) in Cuba. Lima, Peru: Intl. Potato Center (CIP).

McDonald, S.A. 2001. Epidemiology, aphid vectors, impact and management of tobacco etch potyvirus in hot peppers in Jamaica. Ph.D. Dissertation, Virginia Polytechnic Institute and State University.

McDonald, S.A., S.E. Halbert, S.A. Tolin, and B. Nault. 2003. Seasonal abundance and diversity of aphids. (*Homoptera: Aphididae*) in a pepper production region in Jamaica. *Environmental Entomolology* 32: 499-509.

McDonald, S.A., S.A. Tolin, and B. Nault. 2004a. Effects of timing of TEV infections on growth, yield and quality of 'Scotch Bonnet' pepper (*Capsicum chinense* Jacquin) fruit in Jamaica. *Tropical Agriculture* (accepted for publication).

McDonald, S.A., B. Nault, and S.A. Tolin. 2004b. Efficacy of Stylet-Oil® in suppressing the of spread of TEV infections in a 'Scotch Bonnet' pepper (*Capsicum chinense* Jacquin) field in Jamaica. *Tropical Agriculture* (accepted for publication).

McGlashan, D.H., J.E. Polston, and D.N. Maynard. 1993. A survey of viruses affecting Jamaican 'Scotch Bonnet' pepper *Capsicum chinense* (Jacquin). *Proc. Internat. Soc. Tropical Agriculture* 37: 25-30.

Myers, L.R.S. 1996. The etiology of viruses affecting pepper (*Capsicum* spp.) in Jamaica. Masters Thesis, University of the West Indies.

National Research Council. 1992. *Neem, a tree for solving global problems.* Washington, D.C.: National Academy Press.

Nitzany, F.E., H. Geisenberg, and B. Koch. 1964. Tests for the protection of cucumbers from a whitefly-borne virus. *Phytopathology* 54:1059-1061.

Ogrodowczyk, J. 1998. Policies affecting production practices and adoption of integrated pest management for Jamaican farmers in Ebony Park, Clarendon, M.S. Thesis, Virginia Polytechnic Institute and State University.

Palaniswamy, M.S., N. Mohandas, and A. Visalakshi. 1992. An integrated package for sweetpotato weevil (*Cylas formicarius* F.) management. *J. Root Crops* 18: 113-119.

Peterson, A. 1960. *Larvae of Insects, An Introduction to Nearctic's Species, Part II: Coleoptera, Diptera, Neuroptera, Siphonaptera, Mecoptera, Trichoptera.* Ann Arbor, Mich.: Edward Brothers, Inc.

Reid, R. and H. Graham. 1997. *Market research – callaloo, hot pepper, and sweetpotato.* 47 pp. Working Paper for the Agribusiness Council, Kingston Jamaica.

Schlosser, G. 1998. Gendered production roles and integrated pest management in three farming communities, M.S. Thesis, Virginia Polytechnic Institute and State University.

Schlosser, T. 1998. Local realities and structural constraints of agricultural health: pesticide poisoning of Jamaican small-holders, M.S. Thesis, Virginia Polytechnic Institute and State University.

Simons, J.N., J.E. Simons, and J.L. Simons. 1995. JMS Stylet-Oil User Guide: as a fungicide, as an insecticide and for plant virus control. Version 2.1. JMS Flower Farms Inc. 41 p.

Smit, N.E.J.M., and B. Odongo. 1997. Integrated pest management for sweet potato in eastern Africa. Pp. 191-197 in Progress Report, 1995-1996. Lima, Peru: Intl. Potato Center.

Talekar, N.S. 1998. *How to control sweetpotato weevil: A practical IPM approach.* AVRDC Publ. 88-292, Taiwan.

Talekar, N. S. 1991. Integrated control of *Cylas formicarius.* Pp. 139-156 in *Sweet Potato Pest Management: A Global Perspective,* ed. R.K. Jansson and K.V. Raman. Boulder, Colo.: Westview Press, Inc.

Williams, R. 2001. Application of spatial analysis in the incidence of the gall midge in Jamaican hot pepper production. M.S. thesis, Virginia Polytechnic Institute and State University.

— 7 —

Developing IPM Packages in Eastern Europe: Participatory IPM Research in Albanian Olives

Douglas G. Pfeiffer, Josef Tedeschini, Lefter Daku, Myzejen Hasani, Rexhep Uka, Brunhilda Stamo, and Bardhosh Ferraj

Introduction

With the disintegration of the Soviet Union and its transition, along with many other eastern European countries, to capitalism, agricultural sectors were forced to undergo major transformations. Albanian agriculture was no exception, as it attempted to restructure from large collective farms under communism to smaller privately held farm units while merging into a global market using modern practices. In 1991, Albania began to open up to Western countries, although the transition to a market economy has been slow and difficult. Many institutions and modes of interaction and transaction are undergoing drastic changes. In agriculture, production dropped in the mid-1990s due to management problems and lack of inputs. This drop was partly due to problems with privatization, as farmers were not clear about ownership of land or trees, and thus were not motivated to care for their property. This ownership problem has partly been solved, so the farmers are more willing to take better care of the land and trees. About 20% of Albanian farmers are now commercial, and the rest are subsistence. In recent years, the land has been more intensively planted, and some observers have noticed a shift in farmers' attitudes; they are beginning to really understand that no one is going to bail them out, and they must do it on their own.

When land was privatized in the early 1990s, allotment size per family varied. With differing average land size and degree of fragmentation, some districts have varying amounts of land available for agricultural efforts for each family. As a result, there is regional variation in the number of families with a commitment to the land and with plans to continue in agriculture. In a recent USAID-funded survey of farm families, the regions of Lushnja and Korç had the greatest percentages of families planning to maintain land in agriculture (73% and 50%, respectively) (Lemel, 2000).

The Setting

Albania is roughly the size of the state of Maryland. About 75% of the country is mountainous; the remaining 25% consists of a plain from the capital Tirana to the coast. The western coast faces the Adriatic and Ionian Seas. To the north is Montenegro and Serbia, to the east is Kosovo (Serbia) and Macedonia, and to the south is Greece.

All of Albanian public life has been heavily impacted by politics in recent years, with important repercussions for agriculture. The early part of the 20th century showed the rise of nationalism following the collapse of the Ottoman Empire. In the 1930s, a self-proclaimed monarch, King Zog, aligned the country with Italy. Fighting the Fascists and then the Nazis during WWII, the Communist partisan Enver Hoxha was supported by the British and Americans. However, following the war this regime allied itself with Stalinist Russia, later Maoist China, and eventually became extremely isolationist and one of the most repressive dictatorships in history.

The Communist regime fell in the early 1990s. Following widespread public investment in a pyramid scheme, the economy collapsed in 1997, leading to massive rioting and destruction of public and private property. The country is still recovering from this setback, and there have been few resources for farmers to invest in their farms. Privatization of farmland began in 1991. The country was also impacted by the war in neighboring Kosovo in 1999.

Within agriculture in Albania, olive occupies a unique role. It is a major crop: 60% of the population grows olives, and 18 of the 36 districts in Albania have olive trees. National production averages 22,000 tons per year, but it can reach up to 35,000 tons annually. The Berat district leads the country in olive trees, followed by the Vlore region. The Mallakaster region leads in production, followed by Vlore (Ministry of Agriculture and Food, 1997). Olive oil production is 4000 tons per year. The main insect pests of olive are: olive fruit fly (*Bactrocera oleae* [Gmelin], formerly *Dacus oleae*), Mediterranean black scale (*Saissetia oleae* Olivier), and olive moth (*Prays oleae* Bernard).

The level of acidity in olive oil determines the quality rating of the oil: *extra virgin* has less than 1% oleic acid, *virgin* has 1-3% oleic acid, and *virgin lampante* has greater than 3%. While the fruits are still in the trees the level is less than 1%, but after harvest it can rise. Some olive cultivars in Albania have an oil content of up to 30%. There is great interest in improving oil quality. Production of oil is presently 1000-1200 liters per ha but, with proper crop management techniques, this level could be doubled.

With good management a tree can produce up to 24 kg of olives. After 6 years, trees reach full production and, with new micro-propagation techniques, trees can begin producing after 3 years. There are more than 20 olive cultivars in Albania. 3.4 million olive trees were in production in the country in 1996. In the same year 27,660 tons of olives were produced. Yield was 9 kg per tree (Statistical Yearbook, 1996), so there is plenty of potential for yield improvements.

Infestation by the olive fruit fly, *B. oleae*, is a major cause of higher acidity in olives, and lowers the quality of the product. Promptness in processing after harvest is very important since the fruit continues to degrade with time. However, there are some places in Albania where *B. oleae* is not a problem. Economic thresholds are set at 10% infestation for processing olives, and 5–8% for table olives. Without pesticide treatments, over 50% of olive fruits are infested by olive fruit fly, sometimes more than 90%. There is normally one larva per olive. Olives are normally at 1–3 acidity units; with olive fruit-fly infestation acidity rises to 5–7.

There are three generations of olive fruit fly annually. Flies over-winter as puparia. The range of this species is limited by hot summer temperatures; in Albania, the hot, arid conditions of early to mid summer also limit population development. The third generation, shortly before harvest, and occurring as temperatures decline and fall rains resume, may reach very high levels, despite having been at low levels during the first two generations. The female fly deposits eggs singly beneath the skin of olive fruits. Larvae feed internally within the olive for a period of 9–37 days, depending on temperature. Following this period, mature larvae drop to the ground, where they form a pupal chamber from the last larval skin.

There are three generations of olive moth as well. Each brood exhibits a different feeding behavior. Larvae of the first generation (anthophagous) feed on olive blooms and buds. Second generation larvae (carpophagous) feed on more mature olive fruit. Larvae of the third generation (phyllophagous) feed on leaf tissues. The first two generations cause the most economic loss.

Mediterranean black scale, a coccid or soft scale, is another major pest. Members of this family of scale insects generally cause their injury by the copious amounts of honeydew, or undigested sap, produced during feeding. This honeydew supports the growth of sooty mold, which reduces the value of the fruit. Species in this family are also excellent candidates for biological control, since a wide array of hymenopteran parasites attack these scales. In Albania, since growers have not been able to afford to spray their trees for some years now, there is an effective community of parasites in place. A central goal of IPM in olives must be to develop means of controlling olive fruit fly and olive moth that are not disruptive to the naturally occurring control that has developed for Mediterranean black scale.

Diseases on olives include leaf spot, *Cycloconium oleaginum* Cast.; olive knot (*Pseudomonas syringae* pv *savastonoi*); macrophoma (*Spilocoea oleagina*), which invades oviposition punctures of olive fruit fly with the aid of another insect, the olive midge, with high infestation levels when humidity is high; and Verticillium wilt (*Verticillium dahlia*). Olive knot is a bacterial disease that causes galled tissue (parenchymatic proliferations, or knots) on woody growth of olive trees. The bacteria invade the tissue through small wounds associated with harvest operations, hail, or other mechanical sources. Olive shoots may be killed. Olive leaf spot is a disease that causes round lesions on leaves, followed by defoliation.

Identifying and Prioritizing IPM Problems and Systems

In 1998, virtually no IPM was being practiced in Albania. The IPM CRSP, with support from the USAID mission in Tirana, began work in February of that year. A survey of, and discussion among, stakeholders in Albania placed the highest priority on olive, tomatoes, and cucumbers as crops to be addressed by the IPM program. Olive was selected as the crop on which to place the earliest focus, with the suggestion that greenhouse tomatoes and cucumbers be added when funds became available.

Participatory Appraisal

In July 1998, a participatory appraisal (PA) was held, based in Fier, with visits to many olive growers in sections of southern and central Albania, namely Vlore (villages of Kanine, Bestrove, Cerkovine), Berat (Kutalli, Otllake, Bilce), and Fier (Patos, Cakran, Vajkan, Damsi). During the PA, information was gathered on current pest management practices, pest status, perceptions about pests, socioeconomic characteristics, role of women in pest management decision-making, influence of government policy on production practices, and so on. Following several days of farm visits, the research team, consisting of scientists from the Fruit Tree Research Institute (FTRI) in Vlore, the Plant Protection Institute (PPI) in Durrës, and the Agricultural University of Tirana (AUT) returned to Tirana to develop and write up a plan of work (Luther et al., 1999). The plan contained six proposed projects:

1. Baseline survey and monitoring of olive pests and their natural enemies
2. Effect of harvest timing on olive fruit fly infestation and olive oil yields and quality
3. Organic methods of vegetation management and olive insect control
4. Effect of pruning on olive production, infestation by black scale, and the incidence of olive knot and timing of copper sprays to control leaf spot and olive knot
5. Pheromone-based IPM in olive and effects on non-target species
6. Socio-economic analyses

After a difficult period of more than a year, beginning two weeks after the PA (the country was closed to American scientists for more than a year because of threats against the U.S. embassy followed by the Kosovar conflict), three years of field research were carried out.

Baseline Survey

A baseline survey was conducted during the first year of the project. An Albanian member of the IPM CRSP team was a doctoral student in the Department of Agricultural and Applied Economics at Virginia Tech. Given his nationality, he was the only person from the U.S. side of the project able to travel to Albania during the period of American travel restrictions.

The survey was implemented in January 1999 by a team of 14 Albanian research scientists: 4 from the Agricultural University of Tirana (AUT), 3 from the Plant Protection Institute (PPI), 6 from the Fruit Tree Research Institute (FTRI), and the Albanian graduate student from Virginia Tech. Initially, the interviewing team was divided into five groups with specialists from each institution. The mixture of the teams was changed everyday to avoid "enumerator or group biases" during the interviews.

The questionnaire itself was developed based on a previous version that had been used to study pest management practices in rice and vegetables in the Philippines in 1994 (see Chapter 3). Most of the questions had to be modified to fit the specific features of the olive production system in Albania. Additional questions concerning the marketing, credit, institutional, and informational constraints faced by the olive growers in the study area were added.

The survey questionnaire was divided into eight sections. It was translated into Albanian and was pre-tested on ten farmers to ensure that each question fit reality and was understandable by farmers. After the pre-testing, the study team discussed the questions that needed to be dropped, added, or modified in order to come up with the final version of the questionnaire.

The first section of the survey requested background information on olive production systems. The second asked for the pest-management decision making within the household regarding the purchase of pesticides, pest control, and marketing of agricultural products. The third section dealt with factors affecting pesticide use. The farmers were asked to give information about the level of olive production, olive yields, the olive-oil marketable surplus, olive processing technologies, transport and processing expenditures, marketing of olive oil, and the constraints they face. Questions also were included about the amount spent for purchasing pesticides, the importance of various factors affecting their choice of pesticides for different crops, borrowing and credit opportunities, and constraints they face for getting credit.

In the fourth section, farmers were asked their opinions and perceptions about the effects of pesticides on human health and the environment. Section five dealt with farmers' knowledge about olive pests and their natural enemies. Farmers were asked about the olive pests they can recognize, the nature of the damage those pests cause on olive trees and fruits, the natural enemies of olive pests and their respective roles, and their opinion on effects of pesticides on these natural enemies. Section six asked for information about olive pest-management practices including the olive pests encountered during the last season, methods applied for controlling those pests, use of pesticides, timing of pesticide application, the effects of pesticides on those pests, and spraying equipment. Farmers also were asked to list the reasons for not applying pesticides on olive trees.

Section seven dealt with sources of information used by farmers in making decisions with respect to olive pest control. Farmers were asked to identify the most important sources of pest-control advice, as well as about their participation in training courses. Questions were also included about the adequacy of information they receive and the innovative cultural, production, and pest-control practices that they have introduced into the olive-production system. The eighth section asked about the farmer's socioeconomic characteristics including education, years of experience working with olives, tenure status, age, household size, membership in farm organization, and major income sources.

Two hundred farmers were interviewed in five villages, in January 1999. Out of the 200 farm households, 50 households were targeted for interviews with both male and female household heads to obtain gender-differentiated data on roles in pest-management decision making. Overall, 250 questionnaires were completed.

Results of the baseline survey are presented in Daku et al. (2000), with a brief summary of key observations, are summarized below:

a. *Farmer characteristics and practices*: The average farm size was 1.59 ha, ranging from 0.1 to 12.5 ha. The number of trees owned per farmer varied by village. When olive groves were privatized, the number of trees per family depended on the number of people in a village and the number of trees. The number of trees per family ranged from 8 to 200, with an average of 56 trees. Dual-use varieties (oil and table use) made up 65% of the crop, oil varieties 65%, and table varieties only 2%. Olive was the single largest income source for 45% of farmers, and 38.5% rated it as the second largest source. Only about 22% of farmers applied any insecticides to their trees. Among the main reasons for not using pesticides were their cost, uncertainty about product quality from local sources, low crop in some years (olive is subject to biennial bearing), and the fact that neighboring growers don't spray. Some growers use pruning to help control black scale, and soil cultivation to destroy pupae of olive fruit fly.

b. *Farmer knowledge*: The most widely recognized olive pests were olive fruit fly (95.5%), olive psyllid, black scale, olive moth, leaf spot, and olive knot. However, there was considerable disagreement and uncertainly about the nature of damage caused by each pest. For example, 14.5% thought that olive fruit fly damaged branches and leaves. Fewer than 30% of farmers were aware of beneficial insects. Farmers showed little awareness of potential negative effects of pesticides on human health or the environment, although 8% reported that a family member had been poisoned by insecticides.

c. *Trust in information sources*: Most farmers judged their own experience to be the most reliable source of pest control information, usually because there was no other source available. Extension and Department of Agriculture specialists were the second highest regarded source. Pesticide dealers were regarded as the least credible source of information. Specialists at the research institutes were regarded as very reliable, but there was very little contact between research or extension personnel and farmers. Farmers ranked the following criteria for advisors: knowledge, experience, university education, practicality, and accessibility.

d. *Gender analysis results*: The gender analysis completed as part of the baseline survey is discussed in a separate section below.

Monitoring of Olive Pests and Their Natural Enemies

Research on basic phenology of olive pests in Albania provided a basis for further IPM research. Olive moth was sampled using pheromone traps, and by sampling flowers, fruit, and leaves in the respective generations. Data on captures of male and female olive fruit fly were collected on yellow sticky traps. Natural enemies were collected by suction sampling and beating tray. Parasites were reared from olive fruit fly and Mediterranean black scale in rearing cages. Nematodes were collected from soil samples. Leaf spot was assessed on foliage retrieved to the lab, and olive knot by counting knot galls per meter of one-year-old shoots. Weeds were sampled

Figure 7-1. Olive fruit fly monitoring in Albania.

in two 50x50 cm quadrate samples in sample units of five trees, with five collection sites in the experimental grove.

An improved understanding of the phenology of olive insect and disease pests was obtained. Varietal differences were seen: Kokerr Madhi I Beratit (table olive) was more susceptible to olive moth, with 26.5% of the fruit infested, compared with Frantoi (4%) and Kalinjot (1%). The former variety was also more susceptible to olive fruit fly, with 55% of the fruit infested, compared with Frantoi (10%) and Kalinjot (4%). An olive pest new to Albania was found, *Closterotomus trivialis* (Costa) (Miridae: Heteroptera). Greater information is now available on the parasite complex that helps maintain black scale at low densities in Albania. Among the most important parasites is *Scutellista cyanea*, which at times exceeds 80% parasitism; *Metaphycus helvolus* Comp. and *M. flavus* Howard are also important. Parasite identification is important because one of our goals is to develop non-disruptive methods of managing olive fruit fly and olive moth in order to maintain naturally occurring biological control of black scale. Of the parasites of olive fruit fly, the most important was *Eupelmus urozonus* (57% of parasitoids collected), which attacks larvae and pupae of *B. oleae*.

Nine genera and six species of parasitic nematodes were identified from olive trees and nurseries. The most numerous representatives of plant parasitic nematodes were: *Xiphinema pachtaicum, Helicotylenchus vulgaris, H. solani, H. pseudorobustus, Pratylenchus thornei, Gracilacus* spp., *Paratylenchus* spp., *Geocenamus* spp., *Criconemoides* spp., *Tylenchorhynchus* spp. and *Rotylenchulus macrodoratus.* In the root samples, *Rotylenchus macrodoratus* and *Pratylenchus* were found only in the Shamogjin zone, whereas in the olive root in olive nurseries endoparasitic nematodes were not found.

Sixteen different families of weeds were present. Graminacea represented 45.7%, with a few species predominating: *Koeleria gracilis* (L.) Pers, *Poa* sp. (L.), *P. annua* (L.), *Festuca* sp. (L.), and *Alopecurus* sp. (L.). Fifteen broad-leaved families were recorded with Compositae (11.7%) and Leguminosae (16.3%) dominant. The most widespread were *Trifolium* sp. (L.), *Medicago* sp. (L.), *Soncus* sp. (L.), *Conyza* sp. (Lees), *Athyrium filix – femina* (Roth), *Centaurea solstitialis* (L.), and *Polygonum* sp. (L.). Two different kinds of shrubs were present, *Rubus ulmifolius* (Schot) and *Dittrichia viscosa* (L.) W. Greuter.

The highest level of leaf spot infection was observed in May (32.4%). Also, maximum abscission of the leaves was observed a month later, during June (when the temperature increases quickly), and continued progressively during the summer.

Designing and Testing IPM Tactics and Systems

Effect of Harvest Timing on Olive Fly Infestation and Olive Oil Yields and Quality

Research was completed to design several IPM practices. Research in farmers' olive groves found that early harvest of olive fruit may be an effective cultural control tactic for olive fruit fly. As olives mature, oil content and quality increase (Figure 7-2). Depending upon the variety, the normal olive harvest period in Albania falls in late November, the time when olive fruit fly reaches its most damaging population levels (Figure 7-3). During the PA it was thought possible to harvest olives early, before olive fruit fly reached high infestation levels in the fruit. However, data were needed on the olive oil yield and quality when olives are harvested early. Over a three-year period, samples of 100 berries were collected from each of five sample trees harvested at 10-day intervals during September, October, and November. The varieties used Frantoi and Kalinjot, important olive oil-producing varieties in Albania. Oil from these fruit was analyzed at the OLITECN S.R.L. laboratory in Athens, Greece. The research found that early harvesting allows fruit to escape attack by the most intense population peaks of olive fruit fly, while maintaining oil yields. Oil harvest in mid October to early November received a rating of extra virgin; the mid November sample was rated as virgin. Research on this objective provided a cultural IPM tactic that allowed improved fruit quality due to reduced pest infestation, while maintaining acceptable oil yield and quality.

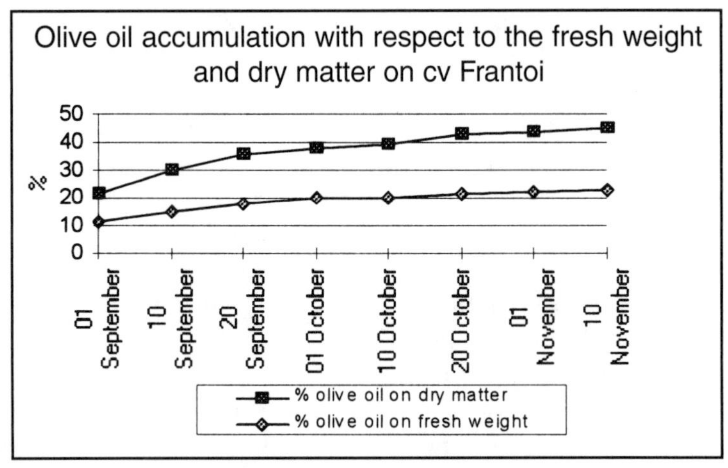

Figure 7-2. Olive oil content of Frantoi olives (note flattening of curve from late October through November).

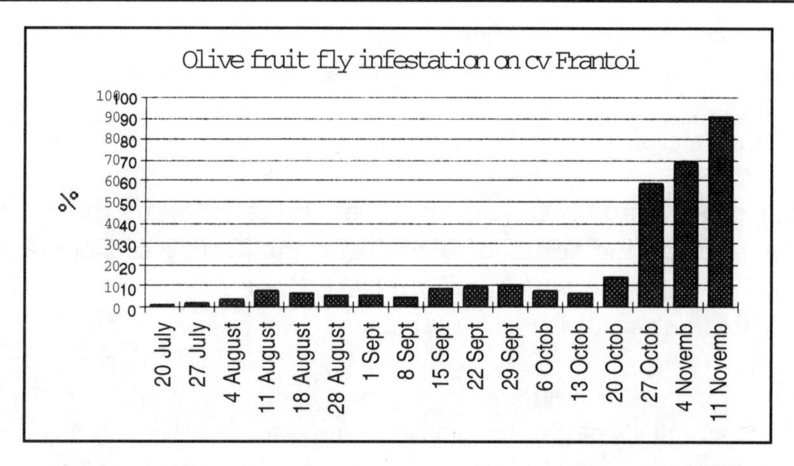

Figure 7-3. Olive fruit fly infestation in olive cultivar Frantoi (note increasing infestation by third generation in late October and November).

Organic Methods of Vegetation Management and Olive Insect Control

The dominant weeds in the experimental field were, for grass species: *Poa* sp. (L), *Cynodon dactylon* (Pers), *Bromus* sp. (L.), and *Agrostis* sp. (L.); for broad leaved species: *Trifolium* sp. (L.), *Soncus* sp. (L.), *Xanthium spinosum* (L.), and *Heliotropium europaeum* (L.); for shrubs: *Rubus ulmifolius* (Scot) and *Dittrichia viscosa* (L.) W. Greuter. While herbicides gave effective control of weeds in olive experiments, straw mulching, an organically acceptable means of weed suppression, also provided good weed control in olive groves (Table 7-1). In the future, mulching treatment should replace the use of herbicides (diuron and glyphosate).

Table 7-1. Total weeds nr/m^2, average and percentage of infestation relative to untreated control (Each stem is considered a weed).

Treatment	R_1	R_2	R_3	R_4	R_5	Average/m2	% of weed
Cover crop	—	—	—	—	—	—	—
Control	294	250	324	342	268	295	100
Glyphosate	12	8	6	4	4	6.8	2.30
Diuron	10	108	94	78	114	80.8	27.40
Grazing	—	—	—	—	—	—	—
Plowing	132	170	230	176	166	174.8	59.20
Straw mulch	46	50	38	36	50	44	14.90

Distinct differences in olive moth susceptibility of olive cultivars were shown. Varietal selection may present another non-insecticidal tool for managing this pest. *Bacillus thuringiensis*, an organically-acceptable microbial insecticide, also provided effective control of olive moth (Table 7-2).

Effect of Pruning on Olive production, Infestation by Black Scale, and the Incidence of Olive Knot and Timing of Copper Sprays to Control Leaf Spot and Olive Knot

Pest effects of three levels of pruning severity (non-pruned, light pruning, and heavy pruning) were tested. Olive trees in the heavily pruned treatment gave good linear vegetative growth. The canopy volume had a good shape, and shoot growth compared with other pruning treatments and non-pruned trees. Also, fruit production was much higher than in the non-pruned and lightly pruned treatment. In addition, water-sensitive papers attached to branches demonstrated that spray penetration is much greater in trees with more open canopies, and the quality of application of plant-protection products could be improved.

Another experiment was conducted applying treatments with copper fungicides every month (October-May) to determine the best moment of spraying to control leaf spot and olive knot (Figure 7-4). The results showed that the treatments during spring (March, April) and autumn (October) are more protective. The management of leaf spot disease was improved effectively.

Table 7-2. Effects of Bt and BI 58 on olive moth larval populations (Anthophagous generation)

Product	Active ingredient	Dose	Observation before treatment 25 May 00		Observation after treatment 1 June 00		Percent mortality
			A	B	A	B	
Dipel PM	BT (16000u l/mg)	0.1 %	10.8	1	2.25	2	79.2 %
(BI 58)	Dimethoate	0.2 %	10.8	1	1.25	1	88.43 %
Untreated plots	—	—	10.8	1	14.5	6.5	—

Note – A = Number of larvae/100 flower; B = Number of pupae/100 flowers.

Figure 7-4. Olive knot infection.

Pheromone-based IPM in Olive and Effects on Non-target Species

An attract-and-kill approach was used against olive fruit fly. The traps used were the Eco Trap type (Vioryl. S.A. Athens, Greece). Traps were 15x20cm envelopes, made of light-green paper, with an internal plastic film lining for water and air proofing. Each trap contained 70g of ammonium bicarbonate salt, a powerful feeding attractant for both sexes, and on its surface 15 mg a.i. of deltamethrin (Decis flow 2.5% AgrEvo Hellas, Athens, Greece), specially formulated for the protection of the active ingredient from natural U.V. light. A pheromone dispenser that was fastened externally contained 80 mg of the major pheromone compound (1.7 – dioxaspiro {5.5} undecane), in racemic form, which acts as a long-range male sex pheromone, as an aggregation pheromone, and as a female aphrodisiac. Eco Traps were placed starting from the end of June at each olive tree. The traps were placed about 2 m high. The Eco Traps were set in olive trees again at the end of August. No traps were installed in trees bearing no fruits. Evaluation of the method was based on the olive fly-population density throughout the experimental period, fruit infestation level, and number of bait spray applications required for acceptable crop-protection levels. Olive fruit fly-population density was monitored using Chromotraps (yellow sticky traps) with sex pheromones. Traps were checked at 5-day intervals. Fruit infestation level, expressed as active infestation (live eggs, L1, L2, L3) and total infestation (live and dead eggs, L1, L2, L3, pupae, and exit holes), was determined by fruit sampling. Fruit damage was assessed every week and the results compared with those where insecticides were applied and with the untreated control continued until the harvest of crop. The samples for fruit infestation level were collected from 10 trees selected at random. From each tree 10 fruits were collected from different points of the canopy. During the course of this research, it was shown that the attract-and-kill approach could significantly lower infestation by olive fruit fly (Figures 7-5, 7-6). However, in years with unusually damp weather, which favors development of the olive fruit fly, supplemental insecticide applications may be needed.

Socio-economic Analyses

An *ex-ante* economic analysis was conducted of the four primary IPM tactics developed for Albanian olives (Daku, 2002). Two alternative starting points were assumed: farmers using a minimum spray program (the most common one currently) and farmers using a full pesticide program. Impacts were projected over a 30-year planning horizon. That analysis concluded

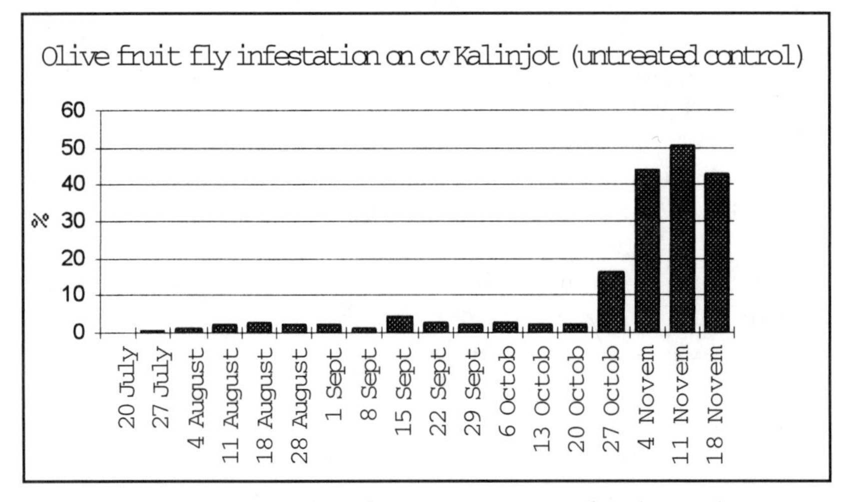

Figure 7-5. Olive fruit fly infestation in untreated Kalinjot olive trees.

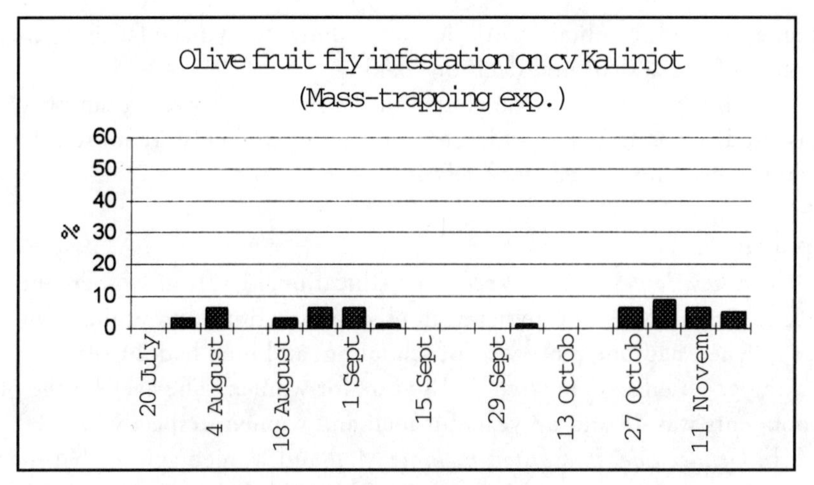

Figure 7-6. Olive fruit fly infestation in pheromone-treated Kalinjot olive trees.

that "...the Albanian olive industry has the potential to derive a net IPM research benefit between $39 million (assuming that farmers move directly from minimum spraying to IPM) and $52 million (assuming that farmers move from full pesticide program to IPM) over the next 30 years" (Daku, 2002). The harvest timing/olive fruit fly study resulted in the greatest gains ($21.1 million), followed by pheromone-based olive fruit fly management ($11 million), weed management ($4.3 million), and pruning/copper sprays ($2.5 million). Of these benefits, nearly 63% were attributed to yield gains, and the rest from quality gains. Consumers accrued 55% of the benefit, and olive producers 45%. Thus, the original 3-year investment by the Albanian USAID mission of $600,000 over three years will have paid excellent dividends to Albania.

Integrating Social, Gender Analysis

Gender analysis is a valuable tool in determining interactions within a community that may influence adoption of IPM results from a research project. There are often gender-based differences in access to IPM educational opportunities, in knowledge of agricultural issues, and in agricultural decision-making. Traditional Albanian society has strictly defined gender roles. It is difficult to assume in advance, however, the effect of such defined roles on IPM adoption. Changes in recent years have been affected by the severe economic situation, which has forced many men to emigrate for extended periods to obtain work, leaving women to manage farms and take on most of the agricultural decision-making.

Of the 200 households included in the Baseline Survey, a sample of 50 households was selected in which both men and women were more extensively surveyed for the gender analysis.

Results

a. *Education/experience levels:* The educational levels of women and men, as well as years of experience in olive production were similar. On average, men had one more year of schooling, and men had 20 years of working experience, compared to 18 years for women. The average age of respondents was 48 and 43 years for men and women, respectively.

b. *Gender roles in decision-making:* Men and women agreed that men played the major role in IPM decision-making, spray practices, and marketing. One difference in responses was that women claimed a larger role for women in marketing decision-making than did the men.

c. *Access to information:* Most men thought that extension specialists were the most credible source of IPM information (where it was available); half of women held this belief, while half felt their own experience was the most credible. However, an extension service is only now under development, so respondents of both sexes had little exposure to extension specialists. Men felt a greater need for training courses; about half of women felt no need for such training.

d. *Knowledge of pests and pest management practices:* Both men and women regarded olive fruit fly as the most serious insect pest of olives. However women were less familiar with other olive pests. Men regarded black scale as the second most severe pest, while women accorded this rank to olive psyllid. Men were somewhat more aware than women of beneficial insects in olive groves, although neither group was highly aware (26% and 18%, respectively). Both groups felt that pesticides are necessary for pest control, but that they are unaffordable and of unknown quality.

Facilitating linkages

At the outset of the project, there was little tradition of cooperative interaction among the three Albanian research institutions. This lack of interaction was fostered partly by lack of resources, and partly by difficulties in travel. There has been progress in this regard over the four years of olive IPM research. Individual scientists have taken the leading role in specific projects, with colleagues from other institutions forming a team.

Other organizations have interacted in sponsoring grower educational meetings. These interactions have strengthened ties among the organizations and growers. These collaborations have supported educational meetings in several olive production districts. Collaborative activities were also carried out with the Interreg Italy-Albania Project. Identification of insects and diseases on Albanian olives was facilitated by the assistance of the University of Bari, Italy. In three workshops organized in collaboration with AATOA, World Learning Project, and the IPM CRSP-Albania, a practical IPM Program for olive was presented and discussed.

Conclusions

This project has developed several non-disruptive tools for use in olive IPM. A cultural method, adjusting harvest timing, may be used to avoid economic losses from olive fruit fly while maintaining acceptable olive oil yield and quality. An "attract and kill" technology was also developed for olive fruit fly. *Bacillus thuringiensis* was shown to be effective against olive moth. All of these methods are compatible with biological control of Mediterranean black scale, so pest outbreaks of this secondary pest will not occur. Straw mulch was shown to be effective for olive weed control, using a resource that Albanian farmers already have on hand.

Economic analysis of the costs and benefits of the olive IPM program developed on the project found that while there was considerable variation in the economic returns among the experiments (harvest timing and olive fruit fly, pruning and timing of copper sprays, vegetation management, and pheromone-based management of olive fruit fly), each of the IPM activities was economically feasible, and all had positive net benefits compared with both current no-spray programs, and a hypothetical full spray program.

The Albanian IPM at a crossroads

The initial funding provided for the IPM CRSP by the Albanian USAID mission was for $600,000 over three years. Because of the year-long period near the beginning when little research could be carried out because of political unrest, it was extended to 4 years. It was planned that olive would be the initial focus of the project, and that work would move into other crops as funds became available. The additional crops originally envisioned were greenhouse tomatoes and cucumbers. Greenhouse crops have a history in Albania, but most of the country's greenhouses are in poor condition, or were destroyed in 1997. There has been considerable interest in developing greenhouse agriculture in the country (Hanafi and Verlodt,

1999; FAO 1999, 2001). Grapes and apples are other crops in need of IPM. Unfortunately the Albanian public sector has been unable to muster sufficient resources, either internally or from donors, to continue the IPM program. Therefore, unlike the other sites on the IPM CRSP, which had USAID (IPM CRSP) funding for a longer period of time that allowed them to become institutionalized in host country programs, IPM in Albania has not been completed to the point where the program is likely to be self-sustaining despite its strong brief success. The lesson is that while excellent research can be completed in a 3-4 year IPM program, a longer-term horizon is required if institutionalization is expected.

Acknowledgments

Other Albanian researchers were critical for the execution and success of the research summarized here, and only editorial restrictions prevented their being listed as authors. They include Mendim Baci, Magdalena Bregasi, Bujar Huqi, Hajri Ismaili, Enver Isufi, Vangjel Jovani, Harallamb Pace, Dhimiter Panajoti, Shpend Shahini, Fadil Thomaj, and Zaim Veshi. Additional participants in the United States are Louise Ferguson, Greg Luther, Milt McGiffen, Charles Pitts, and Beth Teviotdale.

References

Daku, L.S. 2002. Assessing farm-level and aggregate economic impacts of Albanian olive integrated pests management programs: An ex-ante analysis. Ph.D. dissertation, Virginia Tech, Blacksburg. 274pp.

Daku, L., G.W. Norton, D.G. Pfeiffer, G.C. Luther, C.W. Pitts, D.B. Taylor, J. Tedeschini, and R. Uka. 2000. Farmers' knowledge and attitudes towards pesticide use and olive pest management practices in Vlora, Albania: A baseline survey. Working paper 00/1. OIRD/IPM CRSP.

FAO/UN. 1999. Greenhouse vegetable production and protection in Albania. Workshop Proceedings, Tirana. 9-11 March.

FAO. 2001. Mission report, Protected crops in Albania. FAO mission in Albania on protected crops. 10-22 December.

Hanafi, A., and H. Verlodt. 1999. A national plan of action for a sustainable horticulture industry in Albania: A recommended strategy. (Draft). FAO–TCP/Alb/8822 Project: Greenhouse vegetable production in Albania.

Lemel, H., ed. 2000. Involvement in farming and interest in land. Pp. 15-26 in *Rural Property and Economy in Post-Communist Albania*, ed. H. Lemel. New York: Berghahn Books.

Luther, G.C., K.M. Moore, C.W. Pitts, L. Daku, J. Tedeschini, E. Isufi, F. Thomaj, R. Uka, H. Ismaili, B. Teviotdale, D.G. Pfeiffer, M. Bergs, H. Bace, M. McGiffen, L. Ferguson, M. Hasani, D. Toti, B. Stamo, Z. Veshi, and U. Abazi. 1999. Participatory appraisal of olive pest management in Albania to initiate IPM CRSP activities in eastern Europe. Working paper 99/2. OIRD/IPM CRSP.

URL <http://www.cals.vt.edu/ipmcrsp/papers/Alb_wp0299.pdf>

Ministry of Agriculture and Food. 1997. *Statistical Yearbook, 1996*. Tirana, Albania.

Tedeschini, J. 1999. Pest and disease control consultancy. *Greenhouse Vegetable Production in Albania*. Mission Report. Agrarian Studies and Projects Assoc.

— III —
Deploying Strategic IPM Packages

– 8 –
IPM Transfer and Adoption

Edwin G. Rajotte, George W. Norton, Gregory C. Luther,
Victor Barrera, and K. L. Heong

Introduction

Participatory IPM research, through its involvement of farmers, marketing agents, and public agencies, is designed to facilitate diffusion of IPM strategies. However, widespread IPM adoption requires careful attention to a host of factors that can spell the difference between a few hundred farmers adopting IPM locally and millions adopting it over a large area. A number of strategies have been implemented over time in efforts to speed diffusion of IPM around the world. These strategies include working with traditional public extension agencies and approaches and relying on private for-profit and not-for-profit entities that use a variety of specialized training and technology-transfer methods. The complexities of IPM programs; vast differences in local public-extension capabilities; resources, education, and socio-economic differences among farmers; and the need to cost-effectively match IPM strategies to IPM solutions dictates a multi-faceted approach to IPM diffusion if adoption is to be maximized. Given that public resources are scarce, a central issue is how to engage farmers in IPM in a way that maximizes the amount of learning for the resources expended. The purpose of this chapter is to identify some of the lessons learned about how to maximize the depth and breadth of farmer engagement in IPM.

Translating IPM Research Results into Practice

A fundamental difficulty in diffusing IPM knowledge within developing countries is that public-sector extension systems in those countries tend to be weak or non-existent, while the private sector involved in pest management is primarily interested in selling as many pesticides as possible. In areas where farmers can not afford pesticides and are not informed about alternative IPM practices, pest losses are often endemic, and where they can afford them, pesticides tend to be over-used and abused.

Weaknesses in public extension systems have their roots in many factors: low budgets and salaries, poor organization and training, stifling bureaucracies, politicization, inadequate research results to extend, difficulties in distinguishing between public and private goods, and lack of understanding of the potential for extension to accelerate technology diffusion (Anderson and Feder, 2004). In part because of these weaknesses, many public extension systems have shrunk, been eliminated, or been given multiple functions in addition to technology diffusion. In addition, extension duties have been devolved to local governmental units. This devolvement results in a lack of regional and national collaboration in extension programming, and it puts local extension personnel under the authority of locally elected politicians who often divert efforts away from extension programming. In many countries, non-governmental institutions (NGOs) have picked up some extension responsibilities, including IPM, but their programs are usually piecemeal and/or temporary.

The two primary questions that must be addressed in any country hoping to increase the adoption of IPM practices are: (1) which public and private institutional mechanisms can be strengthened and used to speed up the diffusion of IPM knowledge, and (2) what is the optimal mix of approaches for spreading IPM knowledge? Because some IPM knowledge can be conveyed in simple messages (Heong et al., 1998), while other IPM knowledge requires more complex engagement of farmers (Kenmore, 1991), and because of the strengths and weaknesses of various institutional mechanisms, no single approach or institution is likely to be sufficient. Moreover, an understanding of the farmers' socioeconomic situation is imperative in bringing to bear whatever technology transfer entities already exist, or designing new technology transfer systems.

'Supply and Demand' in Technology Transfer

The adoption of new practices by farmers is the result of being provided new knowledge (technology push; supply) and having the new practices demanded of the farmer by others (technology pull; demand). Supply-side factors include the availability of new research information, the appearance of a new farm input, or the presence of an educational program. Demand-side factors include, among others, regulations requiring practice change, competitive advantage in the marketplace for farm products produced using certain practices, and pressure from the local community to change practices.

Any technology transfer plan should address both sides, and as many supply-and-demand elements in given farming systems as possible should be identified in the initial phases of the participatory process. Once identified, each element can be evaluated for its importance in the technology transfer process, and methods can be developed to address each of these elements in the process.

In the Philippines, farmers took annual loans from the Land Bank to establish their rice or vegetable crop. However, the loan from the Land Bank was only half in cash. The other half was provided as inputs such as fertilizer and pesticides. An IPM technology-transfer program that was ignorant of this fact might have tried to promote pesticide reduction, unaware that the farmer might already have obtained a full season's worth of pesticide from his 'loan,' making him reticent to not use the pesticide. On the other hand, this situation also provided an opportunity to change this demand-side factor by educating the Land Bank policy makers about the benefits of IPM so that they would substitute pesticides in the loan program with underwriting IPM practices.

The Importance of Multiple Institutions

Publicly funded extension programs have played important roles in many countries in linking farmers with the latest research knowledge. Public systems can be especially important for reaching small farms and for extending socially beneficial research messages that the private sector may ignore due to lack of a profit incentive. However, some extension systems function better than others and most have become over-extended and less effective where they still exist. As developing countries become increasingly urban and public budgets tighten, public extension services become less valued and their budgets slashed. Therefore as IPM research produces additional knowledge, its broad diffusion is increasingly left to the private sector and to NGOs.

Involving the private sector (cooperatives, export firms, input suppliers, etc.) in IPM diffusion has the advantages that (1) where it is profitable, IPM will be strongly encouraged, (2) use of scarce public resources is minimized, and (3) the demands of the marketplace are brought back to producers. The disadvantages are that the private sector tends to focus on larger farms and on IPM practices that involve products (seeds, chemicals, seedlings, etc.), perhaps to the exclusion of small farms, management-intensive practices, and knowledge that improves the environment but not the bottom line.

Certain types of private-sector involvement can be very helpful in diffusing IPM. For example, cooperatives enable farmers to pool resources and take advantage of technology-transfer mechanisms that might otherwise be difficult to access. Again, in the Philippines, the National Onion Growers Cooperative Marketing Association (NOGROCOMA), whose members produce the majority of the onions grown in the country, support their own technology-transfer agents and test IPM practices, developed with the assistance of public national researchers, on their own demonstration plots and farms. NOGROCOMA is motivated to provide this support primarily because IPM can reduce costs and may offer an advantage when exporting onions to countries that place a value on safely produced food. In Guatemala, fruit and vegetable cooperatives help farmers achieve the quality control required for export markets by transferring IPM knowledge to their growers and developing pre-clearance protocols for export crops.

NGOs often make use of a combination of public and private funds to reach small farms and to address management practices ignored by the private sector. Their grassroots contacts tend to be strong. However, their programs may be targeted to small areas, be of short duration, and have few upstream connections to research knowledge. Because each of these three primary institutional mechanisms (public, private, and NGO) has its strengths and weaknesses, and the relative presence of each differs by country, it will often be the case that a combination of the three is optimal. A responsibility is placed on any participatory IPM research system to involve all three mechanisms for IPM diffusion.

Approaches

Numerous approaches exist for transferring knowledge and engaging farmers in IPM. Within public extension systems, one common approach is to train agricultural extension agents and to station them in villages around the country. The agents are responsible for working with anywhere from a hundred to several thousand farmers. The agents may spend part of their time visiting individual farms in response to problems identified by farmers and part of it holding meetings, demonstrations, and field days with groups of farmers. The group meetings may focus on agendas set by farmers (or youth groups) or may focus on topics identified at higher levels in the extension system (such as occurred under the Training and Visit or T&V system designed by the World Bank). The agents are usually generalists but may be supported by specialists at the regional or national levels. The amount of training they receive and their interactions with research systems will differ by country, as will the supporting materials available such as publications, radio programs, videos, etc. Extension agents may be asked to extend supervised credit or undertake other activities in addition to extending research knowledge.

One common feature of a well-functioning public extension system is that it responds to the demands of farmers, regardless of the commodity or problem. Therefore as new problems arise, they can be addressed relatively quickly or the agents can bring the problem back to researchers for assistance. Because many problems identified by farmers will be pest-oriented, the opportunity for extending IPM solutions is significant, albeit highly dependent on the dedication and training of the agent. The best agents often come from the local areas where they serve and have some knowledge both of the social and cultural environment, and of the farming systems. They reside locally and may be available for consultation on weekends and in informal settings in addition to their structured interactions. Because they reside locally, they also can be held accountable to farmers for the quality of their advice, and the group interactions they have can provide a forum for farmers to share their indigenous knowledge (Figure 8-1).

Because IPM involves attempts to bring all available knowledge to bear on a pest-management problem, and is facilitated by farmer knowledge of

Figure 8-1. Discussing potato IPM with farmers in Ecuador.

beneficial insects, and by farmers experimenting themselves and sharing their knowledge within groups, some public research and extension services, and many NGOs, have taken an intensive IPM farmer-research and knowledge-diffusion approach called the Farmer Field School (FFS) approach. Described in more detail in Chapter 9, the FFS has stressed the importance of farmers growing a healthy crop, observing their fields weekly, conserving natural enemies, experimenting themselves, and using relevant, science-based knowledge. A farmer-training program is held with groups of 12 to 25 farmers. The "field schools" last for an entire growing season in order to take the farmers through all stages of crop development. Little lecturing is done, with farmers' observations and analyses in the field a key component. Farmers and trainers discuss IPM philosophy and agro-ecology, and farmers share and generate their own knowledge (Kenmore, 1991; van de Fliert, 1993; Yudelman et al., 1998). The Global IPM facility at the Food and Agriculture Organization (FAO) of the United Nations has been a driving force in the spread of the FFS approach around the world.

The IPM diffusion mechanisms described above have their individual strengths and weaknesses. Village-level extension agents, if properly funded, have the advantage of providing farmers with the opportunity to place IPM in the context of the total agricultural system. IPM instruction is not tied to just one commodity, and farmer groups with whom the agent works can meet together, perhaps every two weeks, over many years. This system worked extraordinarily well for the Colombian Coffee Federation for many years. Its extension agents addressed all commodities, not just coffee, and were in small part responsible for Colombia's success in establishing itself as a high-quality producer in the world market, with fewer pest problems than many neighboring countries. It was able to reach virtually all farmers in the coffee zone of Colombia.

A weakness of the village-agent approach is that is requires substantial and sustained funding over many years. The example of the Colombian Coffee Federation can be difficult to duplicate in areas without a major export crop. The research and extension activities of the coffee federation were funded in part via a small check-off or tax on exports. In other regions of Colombia served by another government extension agency, agents were not supported as well and were not as successful.

Within the village-agent approach are also other variants of extension methods. Demonstration plots strategically placed around the region/country, with field days held at appropriate times during the growing season (see Figure 8-2, page xxi); radio programs, posters, dramas, and campaigns to spread simple messages can be relatively low cost and yet reach large numbers of farmers for specific messages (Heong et al., 1998). Certain components of IPM programs can be diffused through such methods, and where they are appropriate the methods can be very cost-effective in speeding up technology adoption. Sticky traps, bio-control agents, and grafted seedlings for disease control are examples of IPM practices that may lend themselves to this approach. The Colombian Coffee Federation delayed the spread of coffee rust for several years after Brazil and other neighboring countries experienced it due to active use of posters and radio messages. In Vietnam, pesticide use was reduced over 50% in rice in large areas where the simple message: "no spray for the first 40 days on rice" was widely broadcast. A critical question, however, is how to spread more complex messages rapidly and cost effectively.

Not all IPM extension education must be focused on adults. IPM in school programs can be an effective means of reaching future farmers (and consumers) with information that may change their attitudes and behavior at a young age. Mobil IPM teaching laboratories can be used to reach large numbers of school-age children in areas where widespread training of teachers is difficult.

Transferring IPM across Regions

Farmers fail to adopt an IPM (or any) technology for three reasons: (1) it is unavailable, (2) they are unaware of the it, or unaware that it will help them, or (3) it is unsuitable for their farm due to profit, risk, or other reasons. While the second reason is the primary basis for public support of technology-transfer programs, all three reasons can influence the appropriate design of the technology-transfer system and the ease with which IPM technologies spread from one geographical area to another.

Technology Availability

The major reason farmers fail to adopt IPM is that IPM solutions are not available for their specific pest/crop/location even if it is available elsewhere. Because resources are scarce, research priorities are needed that are based on assessments of projected returns to society (economic, environmental, health, or social returns) locally and across wide geographic areas. Potential aggregate benefits as well as benefits to specific groups must be assessed. Farmers are influenced by numerous considerations within the farm household; therefore participatory approaches such as those discussed in this book are required once the broad geographic focus is determined.

Because regional and global spread of IPM strategies is desired on projects such as the IPM CRSP, information on the importance of the crops and pests beyond the local research domains must be factored into the priority setting process.

Availability of IPM information can be a problem if it is developed with little attention to how it will be commercialized. While input markets respond to profitable opportunities, multiplication and sale of a technology require investments and, in many cases, overcoming a series of regulatory hurdles. Bio-control techniques may require mass rearing of beneficial insects, import and distribution of pheromones, or regulatory approval of a virus that controls an insect. Availability of potentially useful biotechnologies is constrained in countries where bio-safety rules are not in place. The implication is that attention to market and regulatory issues must proceed hand-in-hand with development of the IPM technology if it is to be spread.

Many extension services throughout the world currently have extension agents who are constrained by lack of available technologies. While the fault for lack of spread of IPM is often placed at the feet of extension, in many cases there are extension agents out in the villages who would be willing to extend additional IPM knowledge, if it existed. There are NGOs as well who could do more to extend IPM programs if a greater knowledge base of potential solutions existed.

Awareness of Available Technology

Because farmers are numerous and diverse, are broadly dispersed, and become aware of and understand the need for some technologies more readily than others, multiple technology-transfer methods may be required to cost-effectively reach desired audiences with the depth of knowledge required. Issues of (1) appropriate participation (who, how, when), (2) the need for information on benefits and costs of the technology (implying a need for impact assessment), and (3) the relative effectiveness of various technology-transfer methods must be addressed.

No technology transfer method is a silver bullet. Intense training programs, such as the Farmer Field Schools mentioned above, can be very effective with small groups, but costly to multiply to broader audiences (Feder et al., 2003, 2004). On the IPM CRSP, farmer participation is included at three levels: during research prioritization, during the research process, and during the diffusion process. However, as described in earlier chapters, other groups besides farmers and scientists participate. During research prioritization, policy makers, regulatory officials, marketing agents, technology-transfer agents, and others are included in participatory appraisals. Cooperating farmers from several villages are key players in the subsequent research process, which includes most, but not all, research in farmers' fields. Numerous farmers are also surveyed quantitatively and/or qualitatively early in the research process, to assist with research priority setting, and to permit analyses of factors influencing adoption, both near the center of the research and across potential adoption regions within the country.

A key issue is how to combine the technologies and strategies being developed, with the appropriate amount of participation for maximum diffusion, given public-resource (especially financial) constraints for technology transfer. The IPM CRSP has used Farmer Field Schools extensively in most of its sites and found them excellent for: (1) testing integrated sets of technologies (in our case IPM strategies) that have been developed cooperatively among scientists and farmers, (2) training technology-transfer agents such as extension agents and representatives from nongovernmental organizations (NGOs), and (3) imparting an in-depth understanding of the IPM philosophy to defined groups. A well-functioning field school can achieve its purposes for about $40-50 per participant. Unfortunately, the spread of information from farmer participants to neighbors tends to be weak. Therefore the most effective use of field schools may be to train technology-transfer agents rather than farmers. Some of these agents may then be facilitators in field schools where financial support exists or may utilize other diffusion mechanisms. In all cases, use of mass media should be exploited where possible, and demonstrations and field days be used to spread technology to as broad an audience as possible.

Farmers may be aware of a technology but not adopt it because they are unsure of its net benefits. One reason that the field-school approach is attractive is that it helps farmers themselves explore some of the benefits and costs of various technologies. Information on farm-level economic benefits, generated through credible impact assessments, can also help. Many times the IPM strategy most effective in reducing pest incidence in farm-level trials is not the most profitable option for the farmer due to labor costs or other factors.

Packaging of IPM information can take many forms. Booklets, fact sheets, videos, and radio programs can be relatively low-cost options for information transfer at a superficial level, and can be very useful tools in short- or long-term outreach programs or courses. Computer-aided decision tools, while common in more developed countries, are less utilized in developing countries except for transferring information among scientists or extension personnel.

Technology Suitability

Regardless of how strongly it is recommended, adoption of an IPM technology will not occur if it is unsuitable for a specific environment. Profitability or risk may be related to expected level and variability of yield, or input costs, but many agro-ecological, institutional, and personal factors can determine suitability. And, even if a technology is suitable from a farmer's perspective, external costs and benefits may make it less suitable from society's standpoint. Therefore priority-setting exercises, adoption studies, and impact assessments that focus on only one or a few of these factors will be incomplete.

Technology suitability is often not black and white. Farmers may adopt only part of an IPM package, not because they want to jump in slowly, but because only part of it is suitable to their situation. Adoption and impact-assessment studies in IPM have had to focus on defining IPM adoption first, as adoption is frequently a matter of degree.

Assessing Adoption

Several empirical studies of IPM adoption have been conducted to assess which specific factors or characteristics correlate with adoption decisions by farmers (Napit et al., 1988; Harper et al., 1990; McNamara et al., 1991; Fernandez-Cornejo et al., 1994). The purpose of the studies has been to generate knowledge that will improve technology-transfer approaches and to help predict adoption of new technologies for *ex ante* impact assessment. Most adoption studies use cross-sectional data from farm-households, including them in logit or probit models. Because IPM adoption is often a matter of degree rather than either/or, a simple binary (adopt or not adopt) decision model may not suffice, and the multinomial logit or probit is used with multiple, but discrete, choices to adopt various sets of IPM practices.

A typical IPM adoption analysis uses a logit model in which the dependent variable takes a value of 1 if there is adoption and 0 otherwise. In the binary model, the probability of adoption by the ith farmer is given by $P_i = F(\mathbf{B}'\mathbf{X}) = 1/(1+\exp(-\mathbf{B}'\mathbf{X}))$, where F is the cumulative distribution function (Maddala, 1988). The log-likelihood function of the general multinomial logit model is $\log L = \Sigma i\ \Sigma j\ Y_{ij} \log P_{ij}$, where Y_{ij} is a dummy variable equal to 1 if the individual i falls into the jth adoption category and

0 otherwise. It is assumed that each producer's objective function contains a non-stochastic portion which equals **B'X**, where **B** is a row vector of parameters and **X** is a column vector of exogenous variables. The model can be estimated using maximum likelihood. The marginal effect of a change in a variable on the probability of selecting a specific level of adoption can be computed using the following equation: $\delta P/\delta x_{ij} = \beta_j P_i(1-P_i)$, where β_j is the initial parameter estimate for independent variable j. Goodness of fit of the model can be measured using the McFadden R^2, which is defined as McFadden $R^2 = 1- (\text{Log } L(\beta_{ML})/\text{Log } L_o)$, where Log L (β_{ML}) and Log L_o are the log-likelihood values of the restricted and unrestricted models respectively. The predictive ability of the model can be judged by the number of correct predictions divided by the number of observations.

Examples of applying this type of model on the IPM CRSP are found in Tjornhom et al. (1997), Cuyno (1999), and Bonabona-Wabbi et al. (2002). Cuyno, for example, estimated a logit model to assess factors affecting willingness to adopt IPM practices on onions in the Central Luzon, Philippines. The purpose in her case was to predict future adoption of three specific IPM technologies in different regions. The following categories of variables were included that potentially might affect suitability or farmer awareness of the technologies: (1) farmer characteristics such as age, education, experience, tenure status, (2) managerial factors such as time spent in off-farm employment and the ratio of pesticide costs to total costs, (3) farm structure factors such as farm size, and the share of onion profits in total farm income, (4) physical location, (5) informational factors such as source of pest-management advice and participation in IPM training, and (6) experiences with health or environmental problems associated with pesticides and use of protective measures against pesticide poisoning. The IPM technologies considered in her analysis were burning of rice hulls to reduce nematode problems, trap cropping with castor, and using Bt and nuclear polyhedrosis virus to reduce armyworms. She conducted a survey of 176 farmers, and while in her analysis many of the above factors were found to significantly influence adoption of the IPM technologies, information variables were particularly so, and previous use of protective measures against pesticide exposure.

Tjornhom et al. used a logit model to assess factors influencing pesticide misuse on onions in the Philippines at the start of the IPM CRSP program in that country. They surveyed more than 300 farmers; their analysis of the data found that educational activities and interpersonal contact are the most important factors influencing pesticide misuse. The need for farmer training and awareness was evident in the reduced instances of pesticide misuse by farmers who attended Farmer Field Schools and by farmers who viewed pesticides as harmful to water quality and natural enemy populations.

Bonabona-Wabbi et al. conducted a logit analysis of adoption of eight IPM technologies on cowpea, sorghum, and groundnuts in Uganda. Low levels of adoption (<25%) were found for five of the technologies; three technologies had high adoption (>75%). Results indicated that farmers' participation in on-farm trial demonstrations, accessing agricultural knowledge through researchers, and prior participation in pest-management training were associated with increased adoption of most IPM practices. Farm size did not affect IPM adoption, suggesting that IPM technologies may be scale neutral. Unlike in the Philippines, farmers' perception of harmful effects of chemicals did not influence farmers' decisions with regard to IPM technology adoption.

In addition to these quantitative technology-adoption analyses, focus group and other qualitative techniques as well as simple surveys can be used to identify critical issues that may influence adoption. Identifying issues associated with gender and labor use in the farm-household are especially important. Focus analyses have been conducted on the IPM CRSP in almost every site and have proven useful both for identifying potential constraints to adoption that might have been missed with a survey, and in facilitating follow-up discussions as to why certain factors are constraints. Small surveys of 50 to 600 farmers have also been used to summarize factors that may influence future adoption or to estimate adoption that has already occurred. In Ecuador, for example, 600 farmers were interviewed to assess their knowledge of plantain pests and use of pest-management practices.

Scientist Training

Diffusion of IPM knowledge requires a critical mass of IPM expertise within the research and extension systems of the country. Due to its multidisciplinary nature, education is needed in a range of disciplines including entomology, plant pathology, weed science, nematology, economics, and other social sciences. Graduate education is expensive; hence development of local capacity is essential to minimize the costs. Agricultural graduate programs in developing countries have grown over the past few years in Africa, Asia, and Latin America, especially for masters-level training. On the IPM CRSP, costs have been kept down for Ph.D. education by supporting "sandwich" type programs in which a semester or two of critical course work is completed in a U.S. university and the remainder of courses and research are completed in the host country, sometimes in conjunction with short-term research at an international agricultural research center (IARC) such as the Asian Vegetable Research and Development Center (AVRDC). This type of program also increases the chances of keeping the scientist in the program after the degree. U.S. universities and IARCs also play a critical role in non-degree short-term training.

Future Challenges

Each IPM program must assess its capabilities for sustaining the development of IPM knowledge and diffusing it in a cost-effective manner. Because the capacity of the extension system to deliver IPM messages varies widely by country, the importance of private groups and NGOs has increased over time. However, where extension services still exist, there may be agents who are eager to spread new useful technologies; thus public extension opportunities should not be discounted too quickly.

Social, economic, institutional, and agro-ecological factors all influence adoption decisions. Geographic information systems (GIS) can be used to help identify common geographic areas where IPM technologies are likely to be appropriate, once the importance of specific constraints to adoption are identified. A broad view of IPM adoption constraints is needed because factors such as gender roles, off-farm employment opportunities, and other indirect factors can have a significant influence on adoption decisions. Adoption of grafted eggplant seedlings to control bacterial wilt is spreading faster in Bangladesh than in the Philippines. It is important to know if the reason is the quality and quantity of the technology-transfer mechanisms (awareness), the cost of labor (suitability), or the effectiveness (availability) of the technology.

Conclusions

The key issue in IPM technology transfer is how to cost effectively spread IPM to millions of farmers around the world in enough depth that they will adopt IPM in an appropriate manner. Farmer Field Schools, which are discussed in the next chapter, are useful for transferring in-depth knowledge of crop ecosystems to farmers, maximizing their ability to make appropriate IPM decisions. However, they are expensive and therefore reach a relatively small proportion of farmers. The programs tend to end when the donor funding ends, much as occurred with the earlier T&V system. One option is to focus the FFS programs on training of trainers so that there is a cadre of facilitators with the more in-depth knowledge. Then these trainers can be supported by various mechanisms while they conduct more abbreviated training programs that include meetings, establishment of demonstration plots, distribution of printed material and videos, and field days. If the trainers are public extension personnel, they must integrate IPM training into the other aspects of their program or the public sector will usually be unable to adequately support them.

Too little effort has gone into the rapidly growing possibilities for mass-media instruction, especially through television and radio. Shows such as radio and TV soap operas are popular in many developing countries. IPM messages can be included in these shows.

Too little attention has been devoted to IPM training in schools. While the latest technologies can be imparted through mass media, demonstrations, field days, and in some cases through FFS, the optimal place to achieve an appreciation of agro-ecology is in primary- and secondary-school classrooms. School science curricula are likely to be the most cost-effective mechanisms for helping a broad spectrum of the population gain an appreciation for differences among pests, beneficial insects, and other insects. Once they have this appreciation, their ability to take advantage of short-term IPM messages about the latest technologies and their ability to innovate on their own should be enhanced.

References

Anderson, J., and G. Feder. 2004. Agricultural Extension: Good Intentions and Hard Realities. *The World Bank Research Observer* 19: 41-60.

Bonabona-Wabbi, J. 2002. Assessing factors affecting adoption of agricultural technologies: The case of Integrated Pest Management (IPM) in Kumi District, Eastern Uganda. M.S. thesis, Virginia Tech.

Cuyno, L.C.M. 1999. An economic evaluation of the health and environmental benefits of the IPM program (IPM CRSP) in the Philippines. Ph.D. thesis, Virginia Tech.

Feder, G., R. Murgai, and J. Quizon. 2004. Sending farmers back to school. *Review of Agricultural Economics* 26 (February): 45-62.

Feder, G., R. Murgai, and J.B. Quizon. 2004. The acquisition and diffusion of knowledge: The case of pest management training in Farmer Field Schools, Indonesia. *Journal of Agricultural Economics* 26(July): in press.

Fernandez-Cornejo, J., E.D. Beach, and W.Y. Huang. 1994. Adoption of Integrated Pest Management technologies by vegetable growers in Florida, Michigan, and Texas. *Journal of Agricultural and Applied Economics* 26(July): 158-172.

Harper, J.K., M.E. Rister, James W. Mjelde, Bastiaan M. Drees, and Michael O.Way. 1990. Factors influencing the adoption of insect management technology. *American Journal of Agricultural Economics* 72(November): 997-1005.

Heong, K.L., M.M. Escalada, N.H. Huan, and V. Mai. 1998. Use of commmunication media in changing rice farmers' pest management in the Mekong Delta, Vietnam. *Crop Protection* 17: 413-425.

Kenmore, P. 1991. Indonesia's Integrated Pest Management: A model for Asia. Manila, Philippines: FAO Intercountry IPC Rice Program.

Maddala, G.S. 1988. *Introduction to Econometrics.* New York: Macmillan Publishing Company.

McNamara, K.T., M.E. Wetzstein, and G. K. Douce. 1991. Factors affecting peanut producer adoption of Integrated Pest Management. *Review of Agricultural Economics.* 13(January): 129-139.

Napit, K.B., G.W. Norton, R.F. Kazmierczak, and E.G. Rajotte. 1988. Economic impacts of extension Integrated Pest Management programs in several states. *Journal of Economic Entomology* 81(February): 251-256.

Tjornhom, J.D., G.W. Norton, K.L. Heong, N.S. Talekar, and V.P. Gapud. 1997. Determinants of pesticide misuse in Philippine onion production. *Philippine Entomologist* 11(October): 139-149.

Van de Fliert, E. 1993. Integrated Pest Management: Farmer Field Schools generate sustainable practices, a case study in Central Java evaluating IPM training. Wageningen, The Netherlands: Agricultural University of Wageningen, Thesis 93-3.

Yudelman, M., A. Ratta, and D. Nygaard. 1998. Pest management and food production: Looking to the future. Food, Agriculture, and the Environment Discussion Paper 25, International Food Policy Research Institute, Washington, D.C.

— 9 —
Developments and Innovations in Farmer Field Schools and the Training of Trainers

Gregory C. Luther, Colette Harris, Stephen Sherwood, Kevin Gallagher, James Mangan, and Kadiatou Touré Gamby

Farmer Field Schools (FFS) began in Yogyakarta, Indonesia, in 1989 as a means of facilitating learning of integrated pest management (IPM) concepts and techniques by Indonesian farmers. Since then, many FFS innovations have occurred. The Indonesian National IPM Program first applied the approach on a broad scale with rice farmers, with technical assistance provided by the United Nations Food and Agriculture Organization (FAO) and funding by the U.S. Agency for International Development (USAID) among others. FFS were subsequently adapted for other crops such as legumes, fruits, vegetables, and tubers, and for technical and social themes such as integrated crop management, community forestry, livestock, water conservation, HIV/AIDS, literacy, advocacy, and democracy (CIP-UPWARD, 2003).

Central to the success of FFS programs is IPM and pedagogical training for those who organize and facilitate the field schools. A successful FFS trainer/facilitator needs skills in managing participatory, discovery-based learning as well as technical knowledge of agro-ecology. This chapter summarizes developments and innovations in FFS in recent years, devoting particular attention to IPM FFS training of trainers (ToT).

The Farmer Field School Process

IPM FFS are initiated with introductory meetings in a community to determine if there is interest in establishing an FFS. Usually, an FFS concentrates on a particular crop in order to focus learning and deepen understanding of principles (e.g., on insect ecology, soil fertility, and production economics) that subsequently can be applied to other crops and the farm production system. If the community decides to implement an FFS, the group establishes selection criteria and identifies a group of 15-30 participants. The facilitator and participants draw up a "learning contract" or "moral contract" that includes the commitments of both the facilitator and the group regarding attendance, materials, the management of resources, the investment of the FFS harvest, and other key issues.

Ideally, the farmers contribute the majority of labor and crop input resources, with the facilitator contributing his or her time, transportation costs, and the basic learning materials needed for the FFS. Once norms are established, the FFS facilitator and participating farmers conduct a participatory rural diagnostic on the crop of interest, and identify priority learning themes, as per the different plant growth and production stages. Often the diagnostic focuses on knowledge gaps; i.e., what farmers don't know, but need to know to improve their agriculture. The results of the participatory diagnostic become the learning curriculum. Next, in synchrony with the upcoming cropping season in their area, the farmers and facilitator plant an "IPM plot" and "farmers' practice plot" of the crop to be studied. For perennial crops, existing fields are similarly designated for these two comparative treatments. The conventional plot is managed according to the farming norms of the community, while the IPM plot is managed according to the results of careful agro-ecosystem analysis in order to achieve common objectives — usually the increase of production by area, the decrease of expenditures on external inputs, and the reduction of use of toxic materials. Often the group identifies priority research themes, such as new varieties, the management of a particular pest, or fertilization regimes, and a series of small research plots are planted near the FFS practice plots.

The core activities of the FFS are weekly/biweekly meetings that run the entire length of the crop season for annual crops. For perennial crops, such as trees, the learning is likewise organized around phenological cycles, with the FFS meeting at different locations where distinct growth stages can be found for learning experiments and management practice. These meetings begin with an "Agro-ecosystem Analysis", in which farmers go to the field to observe the crop, pests, natural enemies, diseases, weeds, soil, effects of weather — in essence, the entire agro-ecosystem. Random locations are sampled, and counts are taken of pests and of beneficial and unknown organisms, plant growth and health are assessed, and data are noted on many aspects of the agro-ecosystem. This sampling is often completed in small groups of 3-5 farmers, and each small group subsequently returns to the meeting area and produces a report of what they observed in the field. This reporting involves drawing images on a large sheet of paper of the crop, pests, natural enemies, diseases, weeds, and other components of the ecosystem relevant to understanding and managing crop health. Data on these components are listed in tables to enable other farmers to understand the results. The small group may write brief comments, particularly about their conclusions regarding immediate action that is needed to maintain crop health.

Each small group reports its findings to the large group, and participants are given the opportunity to ask questions and discuss each small group's analysis. This discussion often involves challenging the findings of other groups to encourage both accuracy and appropriate decisions. These exchanges can become quite lively, with farmers expressing their opinions vehemently. Optimally, the group as a whole reaches a consensus about actions needed to maintain crop health.

Often a group dynamics activity will follow the Agro-ecosystem Analysis. These activities usually reinforce technical learning and help the participants (and facilitator) to become more familiar with one another. Many of these group dynamics exercises have additional purposes of exposing the group to the process of analysis and to the importance of cooperating with other farmers in applying IPM in their fields. Role plays and theater activities have been developed that lead to participants portraying, for example, the behavior of beneficial arthropods or disease-causing pathogens. Such activities help farmers understand the important and complex roles of organisms in the ecosystem. The resulting learning experiences, whether technical or of a group dynamics nature, mutually support important aspects of implementing IPM practice in the community.

A 'special topic' exercise is usually the last activity in a weekly FFS session. These topics are optimally identified during the participatory diagnostic and represent an opportunity for the facilitator to introduce material to the group that the farmers are not likely to encounter through their self-directed experiential learning process. These exercises are used to introduce or demonstrate a number of IPM and plant health-related issues, such as toxicity of pesticides to predators and parasitoids, the existence of microscopic pathogens, or the ability of the plant to compensate for damage done by herbivores.

The facilitator and participants work together throughout the FFS process to achieve the learning objectives. An adept facilitator is able to guide the FFS in productive directions while enabling the participants to shape it in ways that meet their needs and interests. Learning is acquired through directed discovery-based experiments that respond to the principal knowledge gaps identified early on. Because of this learning approach, the facilitator does not have to explain everything; the results speak for themselves. Due to the practical orientation of the teaching method, farmers can become accomplished FFS facilitators. The chief skills required are the ability to lead others in carrying out and analyzing field activities. The field becomes as a much a teacher as the facilitator is.

Many variations on the main theme exist for the process of conducting an FFS. Pontius et al. (2002) provide details on what constitutes a standard model FFS and the components of a typical rice IPM FFS. The format and process of the FFS make it possible to address many community issues that influence IPM, such as gender equity, land tenure, and marketing concerns. The FFS process of engaging in inquiry triggers exploration into social issues that affect IPM, crop health, and overall family and community well-being (Figure 9-1).

Goals of IPM FFS

The goals of an FFS vary based on the needs and priorities of a community. Most immediately, IPM FFSs strive to enable farmers to respond to practical needs, usually pest control and improving crop productivity. However, FFSs address individual and collective matters that are often behind social marginalization and poverty (Pontius et al., 2002). Human resource development is an often-stated goal (CIP-UPWARD, 2003). Strengthening farmers' critical awareness, in particular their knowledge base and capacity to analyze problem contexts, experiment, and negotiate mutually beneficial outcomes, is also a commonly held goal. Social empowerment, that is, the capacity to protect economic and cultural interests, is a frequently stated goal, and it was a fundamental principle that influenced the original design and implementation of FFS. Decades of rural development experience have shown that effectively managing interactive learning processes is central to enabling farmers to escape poverty (Pontius et al., 2002). Experiential, self-directed learning around technical concerns, such as IPM, can contribute to increased creativity, self-reliance, and confidence. For this reason, FFS concerns itself with both technical content and socially effective learning processes.

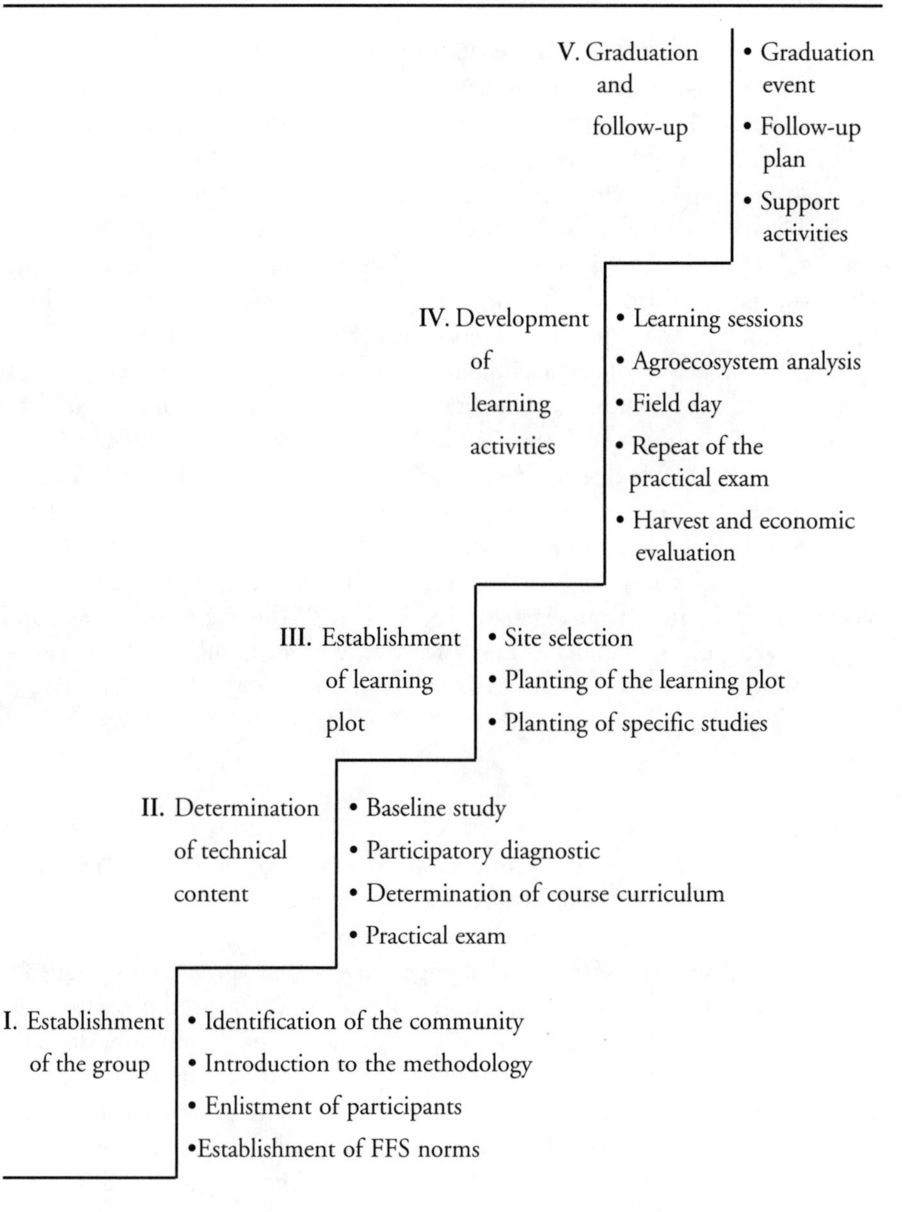

Figure 9-1. FFS methodological process (adapted from Sherwood and Thiele, 2003)

Overall Trends and Global Scope of FFS

Farmer Field Schools are now active in Asia (including East, South-East, South, Central, and Middle East), Africa (Western, Southern, Eastern, and Central), Latin America (South and Central America) and Eastern Europe. FFS-type activities are conducted in Australia on rice and in the United States on fruit trees (OrchardCheck); the basic idea of aligning training with the crop phenology or livestock management and undertaking hands-on practical training has been a "normal" practice in western country organizations such as Future Farmers of America (FFA) and 4-H. The geographic spread has been accompanied by local cultural and socio-economic adaptations by local facilitators. As FFS moved from Asia to Africa, the focus changed from IPM to Integrated Production and Pest Management (IPPM) due to an emphasis on production and already low levels of pesticide use in most crops. Further spread has taken place as the focus of FFS has moved from primarily rice IPM in Asia to vegetable and cotton IPM in Asia to potato IPM in Latin America, cotton, rice and vegetable IPPM in Africa, vegetable and fruit IPPM in the Middle East, and now towards mixed systems in East Africa with crops, poultry, and dairy cows (*LEISA Magazine on Low External Input and Sustainable Agriculture,* 2003; *LEISA Revista de Agroecología,* 2003; CIP-UPWARD, 2003; van den Berg, 2004).

FFSs in Africa

FFSs are presently being conducted by a range of institutions in Africa, including the IPM Collaborative Research Support Program (IPM CRSP), FAO, many national governments, and numerous non-governmental organizations (NGOs). Unique challenges have arisen while attempting to apply in Africa this approach first developed in Asia. For example, efforts to implement FFS in Egypt have found that group dynamics activities developed in Asia do not work in the Arabic-Egyptian culture (van de Pol, 2003). Reorienting FFS facilitators from a top-down technology-transfer approach to a participatory approach has been especially challenging in Egypt, and has required intensive training in the latter over a prolonged period. Overall, adapting the FFS process to local circumstances must be a collaborative activity among farmers, facilitators, and project staff (van de Pol, 2003).

FAO FFS

As mentioned above, the FFSs in Africa focus on production and pest management (IPPM) because of the relatively low levels of production and pesticide usage. Cotton, vegetables, and tobacco are the largest recipients of pesticide treatments. For example, in cotton IPPM, most farmers conclude that they are over-using pesticides and under-using quality seed, irrigation, and fertilizers. In rice IPPM, farmers learn to improve yields without increasing use of (or beginning to use) costly pesticides.

In Africa however, several innovations have taken place since FFSs were introduced from Asia. First is the inclusion of more health and nutrition "special topics" due to the low level of awareness by farmers about the dynamics of diseases such as HIV/AIDS and malaria that are crippling many rural communities. Basic nutrition, water boiling, intestinal parasites, and women's reproductive health are included in FFS by non-IPPM extension officers or NGO guest facilitators. An exciting innovation, which women's groups in Western Kenya developed, are "commercial plots" — group production plots adjacent to the FFS learning plots. Such commercial plots allow the groups to raise funds and become self-financing in their activities. Efforts are underway to institutionalize these commercial plots in the FFS so that they will be largely self-financed from the outset of programs. The International Fund for Agricultural Development (IFAD) is funding a four-country effort to develop the methodology by working with these innovative FFS groups.

The water and soil services of FAO, in collaboration with ICRISAT and national extension, have been especially active in Eastern and Southern Africa developing FFS for soil husbandry, minimum tillage conservation agriculture, soil conservation, water harvesting, and water-moisture management in rain-fed systems. These new field schools combine both educational and participatory technology development (PTD) methods. Further developments for youth (Junior Farmer Field Schools) are also taking place. There are linkages with other FFS developers in the region, especially with IITA in cassava and banana, GTZ in Ghana on banana and vegetables, GTZ in Tanzania on various topics, and with Wageningen University PTD development methods in Zanzibar. A large number of non-FAO programs across Africa have been active contributors to FFS developments.

Also in Africa, FFS are becoming the foundation of field-based food security programs and taking on a new role. Under IPM, farmers learn to better manage their crop for efficient use of resources (time, inputs, etc.). After an FFS, which typically lasts one to two seasons, farmers graduate with new skills. In fact, many groups of farmers in FFSs decide to continue their group as a type of informal or formal association as they have built trust and confidence together. A new trend that is emerging is marketing networks in FFS that cooperate as a larger unit. FFS networks in Western Kenya that consist of about 3000 farmers per district have won supermarket contracts for IPM tomatoes. The skills required for shipping the right quality and quantity at the right time are new to these farmer-owned networks; therefore the FFS curriculum is moving towards management topics as well.

It is critical for FFSs to be able to scale up to more farmers. A program for 250,000 farmers over five years is planned in Sierra Leone, another for more than a million farmers in Kenya, and larger programs in Tanzania. This up-scaling may be possible if farmers are able to lead the largely hands-on activities of a well-designed FFS. The FFS can complement other extension methods including farmer-to-farmer methods that have been used as see-and-do methods for water harvesting and storage. As with PTD methods for production systems, new solutions can emerge from collaboration between farmers and researcher experts; the successful Agricultural Technology and Information Response Initiative (ATIRI) activities by the Kenya Agricultural Research Institute (KARI) is an example. Radio and other mass media play a role for motivation and information exchange, especially where farmer interviews are used.

The IPM CRSP in Mali

As discussed in Chapter 4, the IPM CRSP has regional sites in both West and East Africa, based in Mali and Uganda, respectively. Considerable research has been conducted on green beans, an important export crop in the Bamako area on which exporters are asking the farmers to use large quantities of pesticides. These chemicals are often ineffective, costly, and unnecessary in the Sahelian climate, where green beans can often be produced organically. The IPM CRSP scientists have developed several IPM technologies that can be used for insects and diseases of green beans in Mali, without recourse to pesticides. The main form of transmission of the research results to the farmers is by way of FFS run in collaboration with the Opération Haute Vallée du Niger (OHVN), the local extension service.

IPM CRSP/Mali has been running FFSs for men since the mid 1990s. In 2001 separate women's schools were added in some villages. In spring 2003 a survey of the impact of these schools was carried out in three of the villages — Dialakoroba, Sanambélé, and Tamala — where there are FFSs for both sexes. Twenty-eight female and 24 male FFS members and 30 women and 29 men non-members from these three villages were asked about their opinions of the FFS, what they had learned from it, what they had passed on to others, and which IPM practices they had adopted.

The main issues for the production of green beans in Mali are: (1) thrips, podborer, and whitefly infestations and (2) weeds, particularly where there is no crop rotation. Weed (and soil disease)-control practices include: well-decomposed compost and cabbage inoculated with a biocontrol fungus incorporated into the soil, plastic mulch applied prior to planting for solarization, once-a-day watering, the addition of the *Lonchocarpus* plant to the compost, the use of powdered tobacco, and burning plant residues from the previous season in the plots. Insect pest-control practices include: red Vaseline-covered traps at crop emergence, blue traps at the vegetative stage, yellow traps at flowering, and up to three applications of neem leaf extract (with soap added) if necessary during vegetative growth, at the flower bud stage, and during flowering and pod formation (Gamby et al., 2003). Some integrated crop-management (ICM) techniques, such as planting in rows, were also taught in the FFS.

The aim of the survey was to assess the efficacy of the FFS and the effect of FFS on participants' and non-participants' farming practices, by gender. Questions considered include whether the FFSs can influence views on pesticides and their health and environmental effects and, especially, whether farmers will adopt FFS technologies when the export firms provide pesticides and even a person to apply them along with the seeds.

Although not a statistically valid survey due to the inability to control for selection bias, given the small sample and survey design, and the absence of respondents from non-FFS villages, the results do provide some clues as to the nature of results from FFSs in villages where they have been held in Mali. Perhaps the most significant result is that many participants and non-participants in the villages realize the dangers of applying pesticides, and have therefore decided to attempt to eliminate pesticides from their practices. All but seven of the respondents indicated that they had learned about dangers associated with pesticides during the FFS. The economic benefits of using neem, which grows in the village and can be economical when extraction costs are kept low, also played a role (see Figure 9-2, page xxi). The one person who said he would not be changing his practices indicated it was because he works with the exporters and must follow their rules, despite understanding the dangers.

Most participants and non-participants had adopted several of the technologies and were planning to apply others. Everyone who had practiced any of the technologies claimed to have benefited from them, including economically. Comments included: "The greatest benefit of all is having received training and therefore being equipped with new knowledge and skills"; "The production is better and the quality and yield have improved; there is less danger without pesticide usage; the women in Dialakoroba have formed a group to cultivate cereals together in addition to their green-bean farming; now that we know these principles, we cannot be easily deceived in the matter of production techniques; I can now distinguish between pests and beneficial insects." A majority of women said the chief benefit for them was being able to farm without having to wait for help so they could work to their own rhythms. Some women who had never previously had their own plots, said their new skills had given them the courage to farm independently of their husbands. Many people commented that the FFS had given them the strength to move forward with green-bean cultivation and to cope with the problems.

One perhaps unexpected benefit from the FFS is that when women are able to raise their incomes through use of these techniques, they no longer depend on selling firewood for cash. This freedom from selling wood reduces their workload, since fetching wood is both time-consuming and physically demanding. It may also help the environment.

All participants praised the FFS and said the techniques they had learned were useful, not only with respect to pest management but also for their general farming. FFS had more than met expectations. However, there were several suggestions for improvements, such as more efforts to include youth and women. Attendance would be easier in the dry season when people had more time, and more people could benefit if participants taught those unable to attend. They would like to learn additional technologies and work on other crops.

Most people found the information supplied easy to comprehend. However, the more educated, mainly younger men, felt the lessons could have contained more, while the less educated, such as the older men and particularly the older women, found the amount of material somewhat daunting. The differences are attributable to literacy levels. Those able to write were in an advantageous position, with illiterate participants struggling to remember everything.

The women had clearly learned less than the men, being acquainted with and applying fewer techniques. This result can in part be attributed to sessions missed because of domestic duties. Moreover, their lack of resources meant that they were not always able to apply technologies. This resource constraint was especially notable in the case of the insect traps, which is the most expensive technique to implement.

The main problems that emerged were not FFS-related but involved village conditions, particularly a water problem, which had prevented many people from producing vegetables at all. This lack of water impacted women the most since the men always ensured sufficient water for their own crops before giving their wives a plot to farm.

Many FFS participants had received visits from other farmers or discussed the FFS with family members or neighbors and given them descriptions of the IPM practices they had learned (Table 9-1). These descriptions varied from showing how to use neem leaves and/or apply organic fertilizer to discussing the entire package of IPM techniques. Non-participants appeared interested in learning the techniques, and some asked for help so they could apply them in their own plots. Many participants went over the lessons in detail with spouses and/or neighbors and even showed them practically how to apply the techniques. In cases where several family members participated, they would sometimes hold discussions at home on potential applications to other crops. Many participants were eager for their fellow villagers to learn about IPM to improve their economic levels and reduce pesticide use in the village as a whole.

One comment was that some men had started by mocking the participants for wasting their time, but after hearing about what they were learning and seeing the improvements, the men changed their opinions and were eager to learn more so they also could adopt the new practices.

All but one of the non-participants had heard of the FFS in their village and could name at least one, most of them two to three, IPM practices. It was especially interesting to see the extent of knowledge of non-participants in the village. In July 2001, before the addition of women's FFS, women interviewed in these villages had either not heard of the school,

Table 9-1. Communication of FFS participants with non-participants

	Totals of FFS participants	
Issues	female	male
Received visits from other villagers	24 (86%)	22 (92%)
Discussed the FFS lessons with spouse	14 (50%)	18 (75%)
Discussed the FFS with neighbors	20 (71%)	23 (96%)

or only vaguely knew what it did. Two years later, with the addition of women's schools, non-participant women are not only aware of the schools but also know a great deal about their content.

Almost all interviewees explicitly mentioned the use of neem leaves as an insecticide and of organic manure as a fertilizer, and 37% and 42% of non-participants, respectively, had already adopted them (Table 9-2). 17 others intended to use neem in the near future.

High adoption rates of IPM/ICM practices were found not only for the 52 FFS participants surveyed but also for the 59 non-participants surveyed (Table 9-2).

Fifty-seven out of the 59 non-participant respondents said they thought that the IPM techniques were excellent, that they clearly reduced insects and diseases, and that the beans looked better. Three respondents

Table 9-2. Adoption of IPM/ICM practices by FFS participants (total of 52 surveyed) and by non-participants (59 surveyed)

	FFS participants			Non-participants		
IPM practice	Totals	Women	Men	Totals	Women	Men
mulching	31 (60%)	17	14	18 (31%)	3	15
applications of neem	39 (75%)	19	20	22 (37%)	12	10
well-decomposed organic fertilizer	43 (83%)	21	22	25 (42%)	9	16
colored sticky traps	15 (29%)	5	10	5 (8%)	1	4
tobacco powder	9 (17%)	3	6	0 (0%)	0	0
addition of cabbage residues to fertilizer	12 (23%)	5	7	2 (3%)	0	2
solarization of mulch	10 (19%)	3	7	2 (3%)	0	2
addition of *Lonchocarpus* to fertilizer	3 (6%)	1	2	0 (0%)	0	0
sowing in rows	19 (37%)	7	12	13 (22%)	2	11
burning plant residues	39 (75%)	20	19	18 (31%)	6	12

hoped to join an FFS in the future. Eight people liked the techniques because they enabled them to farm without the use of dangerous pesticides, and 14 commented on the economic advantages of IPM.

Table 9-3 shows considerable adoption of IPM practices by non-participants. The levels would have been even higher, they said, if they had more detailed information about them. Men's adoption rates were higher than women's. The high adoption rates in Tamala may be due to the significant involvement of the FFS trainer with non-participants. Women non-participants had mainly learned the techniques from observing their neighbors, while the men had learned them from similar observations, discussions with IER/OHVN agents, and participation in the original trials. One man even said he believed he knew the techniques well enough to teach them to others. The reasons they gave for adopting IPM practices included: "They are clearly economically beneficial"; "They are safer than pesticides"; "They are obviously superior to the old practices". They came to these conclusions from observing the trial plots and seeing the improvements in the IPM ones and in the fields of their participant neighbors.

Lessons Learned

The results of this survey provide lessons that may be useful to consider when organizing future FFS programs:

- There should be good explanations of the goals and the types of practices to be learned to encourage wider participation. The program should incorporate general farming practices as well as IPM to reduce skepticism by villagers about the FFS being a waste of time.

Table 9-3. Adoption of IPM techniques by non-participants

			Village			
Sex	Non-participants		Dialakoroba	Sanambélé	Tamala	Total
Female	Have adopted at least one IPM technique	No	7	6	1	14
		Yes	2	5	9	16
		Total	9	11	10	30
Male	Have adopted at least one IPM technique	No	5	3	0	8
		Yes	6	7	8	21
		Total	11	10	8	29

- Gender issues are important as women play a large role in farming in Mali. Even where women do not work in the fields they are often exposed to pesticides from washing clothes or reusing empty pesticide cans. Therefore it is crucial to include women in at least some of the training. Where women carry out farm labor, it is crucial to include them in all technology transfer programs, and increasing the numbers of female extension agents would also help.
- The following lessons emerged from this survey in regard to FFS in particular:
  - Women and men have different interests and often have different times when they are available to attend an FFS.
  - There are obstacles to mixed-sex schools in some settings. Issues such as lesson times, cultural acceptance of mixing men and women, women's ability to talk freely in front of men, and women's issues in learning that may differ from men's, especially where they are largely illiterate — all need to be considered. For instance, stricter husbands may refuse to allow their wives to attend a mixed school. Specific cultural obstacles, such as cooking days for African women, need to be avoided to enable women to attend all the classes.
- Extension methodologies:
  - Although most interviewees said they were satisfied with the teaching methodology used in the FFS, those who learn very quickly felt underutilized at times and those who learn very slowly found it difficult to catch up. The illiterate felt they were at a significant disadvantage because they could not write down what the trainer was saying.
  - These problems might be reduced by using more graphic illustrations, fewer written descriptions, and more small-group work in the theoretical sessions. FFS trainers might use tools that give participants a more active role in the theoretical lessons. Using small groups, separated by sex, educational levels, and so on, may also eliminate some of the problems of mixed-sex schools. Those who learn faster can be assigned to support others and made to feel useful as adjunct trainers.

FFSs in Asia

IPM FFSs originated in Asia and has continued to spread to many institutions in the region, including governmental extension programs in various countries and national and international NGOs. Application of the FFS beyond IPM has occurred; it has been applied to community forest management in Nepal (Singh, 2003), gender issues in Indonesia (Fakih, 2003), HIV/AIDS in Cambodia (Yech, 2003), women's self-help groups in India (Tripathi and Wajih, 2003), and a variety of other subjects and areas.

Evolution of FFS in Asian FAO programs and community IPM

The FAO South and South-East Asian Rice IPM Project, coordinated by Peter Kenmore from 1982 to 1997, worked to bring IPM to rice farmers during a period when large pesticide subsidies encouraged over-spraying and release of a secondary pest, rice brown planthopper. This pest caused widespread production losses across Asia. The project focused on removing subsidies for the unneeded rice pesticides as well as promoting farmer education on a large scale. Field training was widely tested and successful in Sri Lanka and the Philippines for farmers and policy makers to understand the role of natural enemies and the disruption caused by pesticides. This training was linked to policy change and — combined with data from national researchers and farmer IPM studies — had a significant impact. The Presidential Instruction by President Suharto in 1986 was perhaps the best known of these changes; it entailed banning 57 pesticides and subsequently removing annual subsidies of US$150 million for rice pesticides. However, policy changes in India, Bangladesh, the Philippines, and other Asian countries also helped to reduce the threat of secondary pest outbreaks.

Large-scale FFS programs emerged first in the case of the Indonesia National IPM Program on Rice, which was later expanded to vegetables and estate crops under various national programs. FFSs were originally designed to fit into the predominant training and visit system, with a few improvements including a hands-on practical field-based curriculum, extension staff as facilitators (rather than being expected to be experts in all fields), and farmer-managed learning plots instead of demonstrations. The learning activities led to successful large-scale implementation of rice IPM. The FFS process has subsequently been adapted to numerous crops and areas in Indonesia.

The Indonesian success was followed by expansion and innovations in Vietnam, the Philippines, Thailand, Bangladesh, India, and China. Malaysia's own practical and hands-on programs informed the overall process, with each participating country contributing its own experiences and improvements. Eventually, the FFSs were no longer only for learning about IPM. Driven by farmer and donor demand for greater sustainability and wider impact, FFSs evolved under the FAO Inter-Country IPM Program towards "community IPM" in which the FFS was adapted for farmers' forums and community associations for focusing on social capital development and environmental, health, and local policy issues related to pesticides and IPM (Pontius et al., 2002). Although many of the "national" projects have not continued after the end of this project, national and local farmers' associations are still active, a testament of the sustainable nature of community IPM. Institutionally, NGOs have taken the place of the FAO programs in many of the countries (e.g., FIELD Indonesia, Srer Khmer in Cambodia).

The IPM for Smallholder Estate Crops Project in Indonesia

The IPM SECP commenced in 1997 and finished in 2004. During this time, thousands of FFSs were conducted following the basic format outlined above. However, there were a few notable differences. Because five of the six main crops in the project were perennials: i.e., cocoa, coffee, black pepper, cashew, and tea, adjustments of the traditional annual crop FFS format were required (Mangan and Mangan, 2003).

The initial attempt to adapt the method of Agro-ecosystem Analysis to tree crops was undertaken by plant pathologists without input from entomologists. The method taught was one of returning to the same sample tree/plant week after week (based on seeing if diseases had progressed). One result was a loss of observational information. It was recommended that the procedure be modified so that participants changed sampled plants every session, and selected them randomly. Another modification occurred after Curriculum Development Specialists noticed that many farmers focused strictly on the limited number of plants they had sampled. There was little flexibility for drawing conclusions from non-sampled plants and other components of the agro-ecosystem. These conclusions could be reached by glancing around as one walked through the plot. For this reason, each FFS was asked to explicitly add a new component, *Pandangan Umum* (General Overview), to their Agro-ecosystem Analysis, to emphasize that farmers need to quickly look over the plot as a whole, and integrate this observation into their conclusions so that they are truly analyzing the agro-ecosystem and not just a small number of sampled plants.

Another problem that resulted from the above tendency to limit sampling to the same spots was that flying insects were largely overlooked. These insects can be very important in the tree-crop ecosystem. Dragonflies and robber flies are important predators, and often one must view them on the wing. Therefore, farmers and facilitators were advised to begin by standing quietly in one spot and watching every flying insect that comes through the space of the tree. After several minutes, the insects observed were entered into the final tally for the group's observation.

Early on in the IPM SECP the use of economic thresholds (ETs) was instituted to help guide farmers with their decisions about when pest-control measures were needed. However, these ETs were being used in a rigid and absolute manner, without considering that ETs can change on a daily basis due to fluctuations in pesticide prices and other factors. The farmers' and trainers' focus on the ET levels often crowded out consideration of other factors they had observed. Farmers were therefore asked to consider a range of factors in the agro-ecosystem when making a pest-management decision, and to keep the ET in mind as a minor factor.

NGOs in Asia

Numerous international and national NGOs in Asia have been conducting FFSs since the 1990s. World Education coordinated and funded a network of Indonesian NGOs to conduct FFS projects beginning in the early 1990s. This network included such NGOs as Gema Desa in Lampung, and Gita Pertiwi and the Institute for Rural Technology Development (LPTP) in Central Java. With small budgets, these NGOs have been able to conduct FFS projects that have produced substantial localized impacts.

LPTP built its program by hiring farmers who were FFS alumni to become full-time FFS facilitators. In addition to training them in participatory methods and technical aspects of IPM, the NGO also facilitated their learning of other new skills, such as how to use computers. LPTP has responded to village needs: in one village where almost all the younger and middle-aged men migrate to the city to work about 10 months of the year and the women do a large share of the farming, LPTP facilitated an all-womens' soybean FFS. Participants ranged from their teens to their 60s. The LPTPs transport FFS alumni to other villages and facilitate discussions among farmers so that useful technologies can spread more quickly.

CARE – Bangladesh has conducted large FFS projects that have trained hundreds of thousands of Bangladeshi farmers since the late 1990s. CARE has included integrated fish culture and rice IPM in the FFS curriculum for its INTERFISH project. Its NO PEST project has also been a large IPM FFS project, focusing on rice and vegetable crops. CARE has attempted to link its training to public agricultural research programs to ensure that FFS facilitators are familiar with the latest research results. It has also linked the FFSs to demonstrations and field days to extend at lest some of the IPM lessons to a broader audience.

FFS in Latin America: The Challenges of Modernization

"Modernization" policies and fiscal constraints throughout Latin America have resulted in the dismembering of agricultural extension and research services. This reduction in support for public extension has transformed the roles of researchers and extension workers and placed greater responsibility on rural communities themselves. Therefore improvements in agricultural research and development necessitate approaches that are responsive and suited to local agro-ecological and socioeconomic conditions. Efforts to introduce FFS have led to a re-thinking of how to organize extension efforts for greater and more effective local agricultural innovation.

Responding to public sector collapse through collaboration

The International Potato Center (CIP), the FAO, the IPM CRSP, and a diverse group of governmental and non-governmental organizations have worked with Andean communities in Ecuador, Peru, and Bolivia to respond to pressing needs of potato farmers. These organizations are striving to enhance farmer understanding of agro-ecosystems and to strengthen local decision-making and technology-development capacities for more productive and sustainable agriculture. Faced with serious pest problems and pesticide abuse, they have emphasized management-intensive approaches that require a strong understanding of biology and ecology.

Beginning in the early 1990s, national and regional research institutes began to work closely with communities to strengthen potato IPM. Building on this experience, Local Agricultural Research Committees (CIALs), developed with the help of CIAT, and Farmer-to-Farmer extension, developed by World Neighbors and others in Central America, have employed a range of participatory extension and research models, in particular FFS methodology.

Researchers engage with communities in collaboration with NGOs and municipal governments. Such collaborative arrangements can yield a variety of benefits. For example, communities gain new access to information and institutional resources, rural development agencies obtain increased technical support, and research organizations gain brokers to mediate between their relatively narrow interests and the broader needs of communities.

In 1997, CIP and its institutional partners in Bolivia and Peru began to experiment with more participatory approaches to training (Torrez et al., 1999a, b), incorporating some elements of the FFS approach, but not Agroecosystem Analysis, which many consider to be its distinguishing feature.

CIP has promoted the FFS approach through a project financed by IFAD (International Fund for Agricultural Development) in six different countries, including Bolivia and Peru. In each country a national research institute and an NGO or other extension organization has been included. In 1999, to support this project, the Global IPM Facility organized a course of three months to train FFS facilitators in Ecuador, Bolivia, and Peru. These facilitators then returned to their work places and implemented FFSs, incorporating other important elements of the Asian model, such as the Agro-ecosystem Analysis. Although many of the fundamental principles have been the same, each country has had its own strategy of implementation, depending on the demands of the farmers and the unique institutional and organizational setting of each context.

In Bolivia, the PROINPA Foundation and the NGO ASAR have taken the lead in designing the training curriculum. Both institutions, in close coordination, have promoted FFSs in different communities. PROINPA has usually taken the responsibility for the research activities and provision of genetic material, and ASAR for the multiplication of seeds of resistant cultivars and the replication of the experience in other places. The main emphasis of these FFSs has been on participatory training. In the learning fields, previously validated strategies of chemical control for late blight with resistant cultivars have been tried (Navia et al., 1995; Navia & Fernández-Northcote, 1996; Fernández-Northcote et al., 1999). Training has concentrated on the use of the strategy and related components. Participatory research activities have been limited to evaluation of new cultivars and advanced clones. PROINPA also supports other related research activities with cultivars resistant to late blight with groups of farmer evaluators, and CIALs composed of farmers (Braun et al., 2000a, b).

In Peru, the NGO CARE has been responsible for implementing FFSs. CIP has taken leadership in developing the training curriculum, in delivering clones and cultivars, and in monitoring the data generated by the farmer-led participatory research. In these FFSs, participatory research has almost the same weight as training (Nelson et al., 2001). The term PR-FFS (Participatory Research - Farmer Field Schools) has been used to represent this approach. Farmers carry out research on the use of cultivars or advanced clones with different degrees of resistance and high, middle, and low intensity of fungicide use, assessing the clones and cultivars by late blight resistance and other qualities. In Peru, the FFSs have also been useful in promoting IPM, in evaluating and disseminating cultivars with resistance, and in generating new information about the efficiency of resistance under different agro-ecological conditions. Here, each FFS lasts for two or three years, with emphasis on research during the first cycle and with a successive transference of responsibility to the farmer group subsequently.

In Ecuador, CIP and INIAP (the national agricultural research institute) have promoted FFSs in the most important potato-producing provinces through a network of local institutions (see Figure 9-3, page xxii). As a result of the recent decentralization, much of the agenda of agricultural development has been placed in the hands of local governments, the NGOs, and the communities themselves. CIP, INIAP, and the Ministry of Agriculture are trying to develop and institutionalize an extension approach based on the farmers and on participatory research methodologies, establishing an effective mechanism of communication between the local institutional actors and the scientists. Here the strategy has been to first increase the local agricultural knowledge through FFSs, and subsequently to support the local process of technological development with participatory research groups such as CIALs, including FFS graduates, research institutions, and universities.

More recently, the FAO established a national FFS program in Peru that has helped in scaling-up IPM in the country. Furthermore, FFSs have spread to Colombia, with the leadership of CORPOICA and FEDEPAPA, and to Central America (El Salvador, Honduras, and Nicaragua) and Mexico, with the leadership of Zamorano/PROMIPAC and the Rockefeller Foundation, respectively. CABI has introduced FFSs to Trinidad and Tobago, and others are testing the methodology in Cuba and Haiti. Within five years of its introduction, FFSs have spread throughout Latin America/Caribbean.

Similar to the African experience, the FFS in Latin America have brought a number of innovations to the methodology as a result of lessons learned in Asia and the unique farming systems and ecologies, institutions, and politics of the region. Thiele et al. (2001) identified specific lessons from the initial experience of introducing FFS to the Andes, some of which are summarized in Box 1. Introducing FFS to Latin America required more than just a re-writing of extension manuals. Partner organizations were generally hesitant to blindly accept external ideas, but they were willing to explore common principles among successful IPM work and to adapt local methods. For example, after agreeing on the benefits of 'discovery learning,' local extension workers took to heart the re-design of their activities to create a new extension guide (see Pumisacho and Sherwood, 2000). The result was an improved approach for the region.

Box 1. Lessons learned from adapting FFS to potato farming in the Andes (Thiele et al., 2001).

1. The economic returns to training is very high in managing late blight.
2. Special care is needed to avoid turning the learning field into a competition between farmers and facilitators or into just a demonstration of the superiority of a new technology.
3. The FFS may play an important role in participatory research, but the roles of other mechanisms and platforms that exist should also be considered.
4. NGOs are valuable partners.
5. Farmers are enthusiastic evaluators of new genotypes, and they do it well.
6. Farmers should take part in trial design or at a minimum the design should facilitate their active participation in trial establishment and data collection.

Presently, the chief challenges to FFS programs are political, institutional, and financial. Impact studies conducted by CIP, INIAP, and the FAO have shown valuable contributions of FFSs to farmer knowledge and a relationship between knowledge and increased productivity (van den Berg, 2004). Other studies in market- and input-intensive areas have shown that FFSs have enabled farmers to significantly decrease dependence on pesticides without negatively harming production per area and in many cases improving overall productivity (Barrera et al., 2001). Despite such impressive results, without public investment in agriculture, it has been difficult for FFSs to reach more than a small group of farmers.

The present challenge for the diverse FFS movements in Latin America is to establish collaborative structures and finance- and technical-support mechanisms to sustain an FFS movement. The diversity of experience has suggested a number of opportunities. For example, in Central America PROMIPAC has tested an IPM labeling system to certify the clean production emerging from FFSs and to link groups to higher-value urban markets. Similarly, groups in Ecuador have established production contracts with the agri-food industry, such as FritoLay and Kentucky Fried Chicken, which provide higher and more stable prices. More work is needed to further develop such market opportunities for FFSs and to coordinate production among groups in order to meet volume demands throughout the year.

Rather than rely on NGOs and professional extension workers that are highly reliant on external funding sources, programs are beginning to work more directly through community-based organizations and are training and supporting local farmers as FFS facilitators. This program shift has led to the exploration of self-financing mechanisms, where the very production of the FFS covers the costs of facilitation. Presently in Ecuador, this modality is beginning to dominate the FFS movement, with the FAO and local governments contributing financial resources to support a small team of technicians and researchers that provides informational and continued training support to farmer facilitators.

Training of Trainers/Facilitators to Prepare Them for FFS

The facilitator is the most important person in the FFS. The success of the entire enterprise depends on having facilitators capable of and willing to encourage participants to direct their own learning processes. Proper training is essential to equip the facilitator to train participants to carry out independent discovery-based learning. This last requirement is crucial if the FFS is to be effective, since those who discover knowledge for themselves tend to use it, while those who are merely provided information do not. Knowledge of the technical side (in relation to pests and beneficial organisms, agronomic requirements for plant health, IPM techniques, etc.) is of course essential to guide the group in productive directions and ensure a rewarding learning experience. If the facilitator does not have command of technical issues, farmers sense that he/she does not know the material and they become frustrated.

Optimally, farmers will actively lead the learning process in the directions they find useful and interesting. However, facilitators must be ready to stimulate those who are unused to the freedom of self-direction, prevent domineering individuals' taking over, and find ways to support the self-development of those who are less forceful. Facilitators also need to know how to work in small groups to allow all to express themselves; this need is particularly important to allow women a voice in mixed-sex schools. Therefore, Training of Trainers (ToT) needs to equip facilitators to tackle a wide range of eventualities.

Training of Trainers and Challenges to Sustaining FFS in Latin America

While diverse organizations, such as CIP and CARE in Peru, had begun to apply aspects of FFS methodology in the late 1990s, it was not until 1999 that the FAO's Global IPM Facility (GIF) conducted the first comprehensive Training of Trainers in Latin America (Sherwood et al., 2000). This intensive three-month activity involved 33 extension workers from rural development agencies and research organizations in Ecuador, Peru, and Bolivia. The thematic platform was the potato, which is a priority food security crop in Andean rural highlands and has major soil-fertility demands and pest problems. Since there was little FFS training expertise in Latin America at that time, the GIF brought in a Master Trainer from Cambodia's National IPM Program, who had previously studied in Cuba and spoke Spanish. CIP and the participants supported technical potato IPM needs. The training content included soil- and plant-health needs of potato growing in the Andes.

Since 1999, FAO and other support has enabled INIAP's National Potato Program, CIP in Ecuador, and the MAG in Peru to conduct a dozen ToTs in FFS methodology, at first in potato and subsequently in a diversity of crops, such as tomato, cotton, and agro-forestry. These programs have trained some 500 professional extension workers from NGOs and farmer promoters. Depending on individual participant needs, programs have included attention to pasture improvement and animal-health concerns. In 2001, a group of Master Trainers from the Andes were sent to support the first ToT in Central America for the Swiss-funded IPM Program for Central America (PROMIPAC) that took place in El Salvador. This training included participants from Honduras, Nicaragua, and El Salvador and centered on field bean, maize, and tomato. Since then, some 200 FFS facilitators have been trained in Central America. The Rockefeller Foundation has supported similar activity in Mexico. This experience was generally presented in a Spanish LEISA Magazine dedicated to FFS in Latin America (see LEISA, 2003).

While ToTs, not unlike FFSs, continue to structure themselves around cropping demands, ToT modality has shifted from intensive 12-week trainings to semi-present, distance educational designs, such as three-day meetings every other week over six months, followed by local implementation of pilot FFSs. This change has been in response to the lack of public support for agricultural extension programs and the difficulties that NGOs face freeing up staff under the pressure of time-bound and objective-driven projects. Funding for ToTs first came from the FAO in South America and COSUDE in Central America. Nevertheless, this support has proven tenuous; increasingly local governments and participants themselves have begun to fund ToTs and FFS in communities.

Most agree that ToTs in FFSs deepen technical IPM competence of regional extension professionals and help to improve participatory methodologies, in particular the use of discovery-learning approaches and group dynamics. In less than five years, FFS methodology was introduced to over 100 rural development agencies in seven Latin American countries that conducted more than 2,000 FFSs in communities. Nevertheless, on-going changes in the region have increasingly marginalized rural development and decimated both public and private capacities to enable rural innovation under modalities such as FFS.

Government restructuring in the region during the 1980s and 1990s, at times encouraged by the World Bank, IDB, and other donors, resulted in non-governmental organizations taking over many rural development responsibilities (see for example, Barrios, 1997; Font and Blanco, 2001). As a result, FFS initiatives in Latin America adopted a multi-institutional, collaborative approach to strengthening individual and collective capacities in IPM and engaged actors from local governments, action agencies, and knowledge-generating organizations. In Ecuador and El Salvador, partners explored direct ties to markets and the food-processing industry. Nevertheless, before alternative funding structures were established, donor agencies began to sharply decrease financing for agricultural development, and NGOs increasingly find it difficult to respond to on-going rural development demands of communities.

While the Swiss Cooperation for Development continues to provide some financing for PROMIPAC in Central America, due to changing organizational priorities and decreasing government capability of supporting national programs, the FAO had by 2003 ended its financial support to FFSs in South America. The public institutional crisis in Ecuador led INIAP to accept funding from CropLife, a pesticide-industry consortium, creating a conflict between public and private interests and leading to a focus on the industry priority of "Safe Use of Pesticides" as opposed to consideration of alternative means of pest management. This situation led farmer organizations and NGOs promoting agro-ecology in the country to call for INIAP to re-evaluate its collaboration with CropLife (CEA, 2004), which it did.

Although municipalities and departmental or provincial governments have been able to fund a handful of ToTs, it is clear that local governments are not capable of maintaining the long-term technical, capacity-building, or coordination support needed to enable coherent IPM FFS programs. While the pesticide industry has occasionally expressed interest in FFS methodology, not surprisingly its interest in FFSs has waned when IPM begins to emphasize pesticide-use reduction and the elimination of problematic products (see BBC, 2004). In the absence of lasting public or private industry support, a number of examples of self-financing mechanisms to fund FFSs at the community level exist. Volunteer farmer promoters usually run these. Nevertheless, without continual attention to broader needs of facilitators, in particular social organization and conducive governmental policies, the FFSs in the region may be in peril, at least in their current form.

Training of Trainers in Asia

China

FAO began to support FFSs in China in 1992 and expanded to the first ToT in Hunan Province in 1993. The Chinese extension system made several innovations necessary to carry out Training of Trainers. First, county-level crop-protection technicians had numerous responsibilities, and their work units were not willing to lend them to a training program for more than a single rice season, if that. This constraint required that the crop IPM training season, during which candidate facilitators learned IPM by doing experiments on a rice crop, be overlapped with the "extension season" where the trained facilitator organized his/her first real FFS in the field. In order to do both kinds of training at once, a system was devised in which several FFSs were organized within the ToT. Candidate facilitators were put through all the activities to be carried out among farmers before FFS day. Each candidate facilitator was given responsibility for a group of five farmers for such activities as Agro-ecosystem Analysis or a defoliation trial. Larger group activities, such as group dynamics, were completed with all farmers together. Even though they did not know the outcome of the experiments or the condition of the ecosystem later in the season, the candidate facilitators knew enough before FFS day to have their small group carry out their Agro-ecosystem Analysis, or set up their trials, on the scheduled day. Only as the season progressed did the candidate facilitators themselves become aware of the purpose of many of the FFS activities through which they guided their small group of farmers in a ToT environment.

Since everyone became aware, before the end of the season, of why they carried out these activities, there was no loss in not explaining the purpose behind the activities before starting them. In fact, there might have been a gain. Discovery can be more powerful than repeating someone else's experiment, or being told beforehand why a field trial is to be completed.

These innovations in scheduling, in which both field activities and FFS farmer training were compressed into a single season, were then used in other countries such as India and Pakistan. The more efficient use of time that this approach made possible helped to promote the process of FFS ToT for such long-season crops as cotton in countries that would otherwise have been unwilling to commit themselves to the long duration of training required.

One study in China (Mangan and Mangan, 1998) compared FFS IPM training with another model of multiple-session IPM training, concentrating not on the whole ecosystem but only on pests and diseases. Interviews were carried out with both groups of IPM trainees before training, immediately after training, and one year later after two rice crops. The most significant finding was that the FFS farmers' learning "took off"; the number of beneficial insects they identified after a year was higher than right after training. On the other hand, the identification skills of the pests and diseases of IPM trainees degenerated after a year, and they became more reliant on using pesticides than on understanding the ecosystem. FFS farmers used significantly less pesticide, but crop yields for both groups were not statistically different.

The IPM SECP in Indonesia

Most FFS facilitators on the IPM SECP initially lacked knowledge of the roles/functions of the wide range of insects and spiders in their crop ecosystems. While knowing the correct names of these organisms is not crucial to implementing IPM, knowing their roles — pest, natural enemy (predator, parasitoid, or pathogen), or neutral species — in the agroecosystem is important. Elucidating their roles through observation in "insect zoos" is recommended, but in many instances time is limited for such observation on a wide variety of organisms, especially when a pest-management decision must be made quickly. Also, the behavior of some insects does not lend itself to easy observation in an insect zoo.

To respond to the need for FFS facilitators to know the insects in their crop ecosystems, a TOT program was designed. Because facilitators in a large number of Indonesian provinces needed training in a short time, participatory activities in the field were prioritized in the curriculum.

One objective of the ToT was to enable FFS facilitators to identify insects/spiders as pests, natural enemies, or neutral organisms. These roles often divide down the lines of Orders and Families within the insect world; for example, all mantids (Order Mantodea) are predators, so if a facilitator/farmer can identify an insect as being a mantid, then s/he would immediately know this is a natural enemy that could help control pests. Similarly, all spiders are predators. We moreover chose an insect classification system that lent itself to this learning need. For example, some systems place Mantodea as a Suborder together with grasshoppers and roaches, all under the Order Orthoptera.

Each ToT began with a presentation to the trainees on major insect Orders and spider Families, using many color photos. Separate presentations were prepared for each crop ecosystem to make them more specific to trainers' needs. Focus was on specific natural-enemy groups that would likely attack pests on that crop. Questions and discussion during the presentation helped make it more participatory.

The trainees then conducted an agro-ecosystem census on their particular crop. They first spent several hours in the field in small groups collecting all arthropods they could find on or near the crop. Then they returned to the work area to categorize them into pests, natural enemies, or neutral organisms based on previous experience and knowledge, use of some written materials in Indonesian (e.g., Program Nasional P&PPHT, 1991), and guidance from consultants. Insects and spiders were identified to the Order level or — if that was not enough to determine which category they fit into — to the Family level. Each small group spent hours assembling and examining their collected specimens and labeling them, and these labels were judged correct or not by consultants. Each small group reported their findings to the others, with discussion among the entire group.

Another activity in the ToT was for each FFS facilitator to find a predator in the field and follow it for one to two hours. Each facilitator made a brief oral report of his/her observations, and much was learned through this process. For example, those who followed dragonflies and robber flies — the two most difficult predators to follow among trees — discovered by compiling their observations that these insects would return to the same branch after each attempt to catch prey. They discovered that these flying insects are territorial — that they do not simply roam randomly throughout the archipelago.

An activity to help trainees understand the food web in their agro-ecosystem was also conducted. Trainees were each assigned a role as one of the pests or natural enemies on their crop, or the crop itself, or the sun, and each drew their organism/role on paper. They then stood for that organism/role in the web. Together with the consultants they constructed a food web by tying colored string from one trainee to another to signify who ate whom. Yellow string was used from the sun to the crop to signify the plant's obtaining its energy from sunlight. Green string was tied from the crop to the pests to signify these herbivores feeding on plant material, and natural enemies feeding on pests was signified with red string. By tying people together, everyone could physically "feel" the relationships in the agro-ecosystem, an enhancement of the learning process. It is hard to forget a food web that you have been tied up in.

The ToT usually wrapped up with a drama activity in which FFS facilitators divided into pairs and received a natural enemy-prey combination from the consultants. Each pair acted out their combination and the rest of the group guessed which natural enemy and prey species they represented.

Training of Trainers in Africa

Training of trainers in Africa has largely followed methods developed in Asia except that computer literacy, HIV/AIDS, nutrition, and other issues have been added to the curriculum. There is an emerging emphasis on marketing networking among farmers; therefore ToTs also include network management and marketing methods. IPM is an excellent entry point for the FFSs but wider issues can be incorporated to make them relevant to farmers and extension services. New efforts in post-conflict situations in Uganda, Congo, Sierra Leone, and Liberia also include non-government extension staff and bring in greater social awareness and peace-building methods. FFSs can play a role in the recovery process of rural institutions, production, and markets in these countries, including further development of youth-based field-school methods.

Conclusions

Each region has its specific obstacles to overcome to successfully implement FFSs, but one that appears to be common to all is the difficulty of reorienting facilitators from a top-down teaching approach to a participatory mode of operation that enables discovery-based learning. This obstacle is especially a problem for government extension agents who may have operated their entire careers in a top-down, "tell the farmer what they should do" mode. Even after training in the participatory approach, the tendency is to revert to top-down teaching as a default, especially when a challenging situation arises. Intensive training in the participatory approach over an extended period of time appears to help with this problem.

The FFS experience reminds us that many solutions to agricultural problems are not technological but conceptual and social. FFS can respond to the needs of an individual farmer as well as the social needs of the community in ways that contribute to farming innovation and productivity.

FFS is a knowledge-intensive learning and action process that emphasizes the management of ecologies, and many professionals, institutions, and organizations are not well positioned for this process. Governments attempt to handle budget crises through privatization of public services or reduced expenditures on public extension, and this financial constraint reduces support for FFS programs. FFSs are intensive but expensive per farmer reached and, as a result, IPM as taught through FFSs continues to reach only a relatively small percentage of rural farmers and communities. There are several options that can be explored to address this issue. First, attention can be devoted to improving the flow of information and technology from FFS participants to non-participants. Second, working with new partners, such as groups based in local communities, may help to expand the financial base of support. Third, developing FFSs for farmer promoters who can then organize and train other groups of farmers should help. Fourth and fifth, further developing self-financing opportunities, and complementing the FFSs with mass-media methods, should help FFS to reach more farmers.

A recent synthesis of 25 impact evaluations of FFSs found "substantial and consistent reductions in pesticide use attributable to the effect of training" and increases in yield were found in many instances (van den Berg, 2004). Although some of these evaluations were subject to methodological flaws, evidence indicates there have been impacts from FFSs, at least locally. Farmer Field Schools have produced positive outcomes, and the basic approach is being refined and more closely linked to other approaches, such as demonstrations and mass-media methods, to reduce the overall costs in extending IPM to large numbers of farmers.

References

Barrera, V., L. Escudero, G. Norton, and S. Sherwood. 2001. Validación y difusión de modelos de manejo integrado de plagas y enfermedades en el cultivo de papa: Una experiencia de capacitación participativa en la provincia de Carchi, Ecuador. *Revista INIAP*. 16: 26-28.

Barrios, F. 1997. El futuro de las ONG's o las ONG's del futuro. Gestión de recursos, proyectos e intervenciones. Bolivia: Plural Editores.

Braun, A., G. Thiele, and M. Fernandez. 2000a. CIALs y ECAs: Plataformas completarias para la innovafcion de los agricultores. *LEISA Boletin de ILEIA*.

Braun, A., G. Thiele, and M. Fernández. 2000b. Complementary platforms for farmer innovation. *ILEIA Newsletter*, July: 33-34.

BBC (British Broadcasting System). 2004. *Dying to make a living. A two-part World Service program on globalization and pesticides*. Aired worldwide in May. For more information see <www.dyingtomakealiving.com>

CEA (Coordinadora Ecuatoriana de Agroecologia). 2004. Porque decir no: Llamada para la soberanía de las agencias públicas y enfoques más agroecológicos en Carchi. Public letter to the Director of INIAP. 6 pp. available at <www.dyingtomakealiving.com>

CIP-UPWARD. 2003. Farmer Field Schools: From IPM to platforms for learning and empowerment. Users' perspectives with agricultural research and development. Los Baños, Laguna, Philippines. 83 pages.

Fakih, M. 2003. Gender field schools. *LEISA Magazine on Low External Input and Sustainable Agriculture* 19(1): 26-27.

Fernández-Northcote, E.N., O. Navia, and A. Gandarillas. 1999. Bases de las estrategias del control químico del tizón tardío de la papa desarrolladas por PROINPA en Bolivia. *Revista Latinoamericana de la Papa* 11: 1-25.

Font, J., and I. Blanco. 2001. Conclusiones. In: *Ciudadano y decisiones públicas*, ed. J. Font. Barcelona. España: Ariel Ciencia Política.

Gamby, K.T., R. Foster, A.T. Thera, K.M. Moore, J. Caldwell, S.H. Traore, and A. Yeboah. 2003. Development of an integrated package for management of diseases and insect pests on green beans. IPM CRSP Ninth Annual Report, 2001-2002. IPM CRSP Management Entity, OIRED/Virginia Tech, Blacksburg, Va. pp. 225-228.

LEISA Magazine on Low External Input and Sustainable Agriculture. 2003. Volume 19 (no.1, March). Leusden, The Netherlands: ILEIA. 36 pp.

LEISA Magazine, 2003. Apriendo de ECAs Volume 19(1, June). 88 pp. Available at <www.leisa.org.pe>

LEISA Revista de Agroecología. June 2003. Volume 19 (1, June). *Asociación Ecología, Tecnología y Cultura en los Andes*. Lima, Peru. 87 pages.

Mangan, J., and M.S. Mangan. 2003. FFS for tree crops. *LEISA Magazine on Low External Input and Sustainable Agriculture* 19 (1, March): 30-31.

Mangan, J., and M.S. Mangan. 1998. A comparison of two IPM training strategies in China: The importance of concepts of the rice ecosystem for sustainable insect pest management. *Agriculture and Human Values* 15: 209-221.

Navia, O., H. Equize, and E.N. Fernández-Northcote. 1995. Estrategias para el control químico del tizón. Fitopatología. *Ficha Técnica* 2. Cochabamba, PROINPA.

Navia, O., and E.N. Fernández-Northcote. 1996. Estrategias para la integración de resistencia y control químico del tizón. Fitopatología. *Ficha Técnica* 3. Cochabamba, Bolivia: PROINPA.

Nelson, R., R. Orrego, O. Ortiz, J. Tenorio, C. Mundt, M. Fredrix, and J. Vinh Vien. 2001. Working with resource poor farmers to manage plant diseases. *Plant Disease* 85: 684-695.

Pontius, J., R. Dilts, and A. Bartlett. 2002. Ten Years of IPM Training in Asia — From Farmer Field School to Community IPM. FAO Community IPM Programme, FAO Regional Office for Asia and the Pacific, Bangkok. 106 pages.

Program Nasional Pelatihan dan Pengembangan Pengendalian Hama Terpadu. 1991. *Kunci Determinasi Serangga.* Yogyakarta, Indonesia: Penerbit Kanisius. 223 pages.

Pumisacho, M., and S. Sherwood (coordinators). 2000. Learning tools for facilitators: Integrated crop management of potato [in Spanish]. INIAP/CIP/IIRR/ FAO. 188 pp.

Sherwood, S. G., R. Nelson, G. Thiele, and O. Ortiz. 2000. Farmer field schools in potato: A new platform for participatory training and research in the Andes. ILEA. December. 6 pp.

Sherwood, S., and G. Thiele. 2003. Facilitar y dejar facilitar. ECAs: Experiencias en American Latina. *LEISA* 19(1): 80-84.

Singh, H. 2003. Community forest management and FFS. *LEISA Magazine on Low External Input and Sustainable Agriculture* 19(1, March): 13-15.

Thiele, G., R. Nelson, O. Ortiz, and S. Sherwood. 2001. Participatory research and training: Ten lessons from Farmer Field Schools in the Andes. *Currents.* Swedish University of Agricultural Sciences. 27 (December). 4-11.

Torrez, R., J. Tenorio, C. Valencia, R. Orrego, O. Ortiz, R. Nelson, and G. Thiele. 1999a. Implementing IPM for late blight in the Andes. *Impact on a Changing World.* Program Report 1997-98. Lima, CIP. Pages 91-99.

Torrez, R., A. Veizaga, E. Macías, M. Salazar, J. Blajos, A. Gandarillas, O. Navia, J. Gabriel, and G. Thiele. 1999b. Capacitación a agricultores en el manejo integrado del tizón de la papa en Cochabamba. Fundación PROINPA.

Tripathi, S., and S. Wajih. 2003. The greening of self help groups. *LEISA Magazine on Low External Input and Sustainable Agriculture* 19(1, March): 24-25.

van de Pol, J. 2003. The Egyptian experience with FFS. *LEISA Magazine on Low External Input and Sustainable Agriculture* 19(1, March): 22-23.

van den Berg, H. 2004. IPM Farmer Field Schools: A synthesis of 25 impact evaluations. Wageningen University, The Netherlands. 53 pages.

Yech, P. 2003. Farmer life schools in Cambodia. *LEISA Magazine on Low External Input and Sustainable Agriculture* 19(1, March): 11-12.

– 10 –
Pesticide and IPM Policy Analysis

George W. Norton, Jessica Tjornhom, Darrell Bosch,
Joseph Ogrodowczyk, Clive Edwards, Takayoshi Yamagiwa,
and Victor Gapud

Introduction

The need for increased food production from limited land resources has fueled an intensification of agriculture and increased reliance on chemical pesticides. Traditional farming systems that rely more on a dynamic equilibrium between pests and beneficial organisms have declined (Shillhorn van Veen et al., 1997). Economic forces such as rising labor rates have encouraged the substitution of low-cost pesticides for labor, but so too have many direct and indirect government policies such as pesticide subsidies, over-valued exchange rates that reduce the cost of imported pesticides, and crop insurance rules. Biological controls, cultural management, and other IPM practices often find it difficult to compete with policy-induced incentives to increase pesticide use. In addition, as knowledge grows about the relative health and environmental risks associated with specific chemistries, policies are needed to regulate or restrict access to chemicals that have harmful effects on public health. Finally, government policies can encourage or discourage the development and diffusion of IPM technologies. Consequently, policy analysis is often an integral part of a successful IPM program.

Upon initiating a participatory IPM program in a specific country or region, policy analysis may begin during the participatory appraisal (PA) when producers and input suppliers are questioned about pest-management practices and institutional constraints and policies that influence their pest-management decisions. Observations are made about the types of pesticides applied (Class I, Class II, Class III), application practices, pesticide prices including direct input subsidies, sources of pest-management information, credit or insurance programs that encourage pesticide use, and so forth.

Issues are identified for more in-depth analysis. Eventual IPM policy prescriptions are likely to involve proactive policies to reduce environmental externalities associated with pesticides and encourage IPM adoption. These policies include development of a regulatory environment to encourage appropriate pesticide use, public support for research and technology diffusion on IPM, economic incentives such as taxing of pesticides and subsidies for IPM practices (instead of the reverse), and information dissemination to increase awareness of pesticide hazards. Therefore, standard types of policy analysis are to: (1) measure the extent of existing economic incentives to apply pesticides or use IPM practices, (2) review existing regulatory mechanisms, and (3) review existing pest-management research and technology diffusion programs.

Economic Incentives

A wide variety of economic incentives may influence the use of pesticides or IPM, but the most common types of policy analyses focus on assessing the extent and impacts of direct pesticide input subsidies, tariff treatment of pesticides, exchange-rate policies that indirectly influence pesticide prices, credit or crop insurance programs that require pesticide use to obtain the loans or insurance, grant or subsidy programs to encourage adoption of IPM, and effects of international agreements that influence residue levels on traded products. Other policies include pesticide sales tax exemptions, cost subsidies to pesticide manufacturers, import restrictions on plant material, and rules with respect to intellectual property rights protection. The approaches used for policy analyses involve combinations of qualitative techniques to trace through the policy process for the purpose of identifying the key decision-makers for specific policies, and quantitative methods to assess both income and environmental effects of direct and indirect distortions on producers, consumers, and manufacturers.

Regulatory Mechanisms

IPM programs are greatly influenced by regulatory mechanisms in place to manage the handling and use of pesticides. Examples of common pesticide regulatory activities are presented in Table 10-1. Most countries have a government agency or authority responsible for regulating pesticides and overseeing environmental protection and/or food safety related to agrochemicals. It is essential for each country to establish an overall pesticide policy and appropriate regulatory mechanisms. Many developing countries take their lead in these matters from the United States and Europe, and from international bodies such as the World Health Organization (WHO) and Food and Agricultural Organization of the United Nations (FAO). The WHO, for example, groups pesticides into classes of toxicity based on oral toxicity to mammals. The FAO has detailed guidelines on procurement, trade, and handling of pesticides under the uniform code of conduct on the Distribution and Use of Pesticides (FAO, 1990). The major difficulty often faced by developing countries is maintaining adequate enforcement of pesticide regulations given the costs involved.

Table 10-1. Common pesticide regulatory activities

Type of Regulation	Enforcement Process	Target Group
Pesticide registration	Quality control and risk assessment of new pesticides	Manufacturers, distributors
Pesticide Re-registration of new formulations	Continuous risk assessment of new agro-chemicals	Manufacturers, distributors
Banning unsafe pesticides	Banning sales with monitoring and penalties for noncompliance	Manufacturers, distributors, end users
Food Safety	Development and monitoring of standards	Producers, processors, researchers
Pesticides in groundwater	Monitoring residue and prosecuting violators	Producers, manufacturers
Pesticide storage and disposal	Certification, inspection, prosecuting violators	Producers, distributors, applicators
User or farm-worker safety	Certification, training, monitoring	Producers, applicators, farm-workers
Resistance prevention	Voluntary or mandatory participation	Producers, distributors, applicators

Source: Shillhorn, Forno, Joffe, Umali-Deininger, and Cooke (1997).

Research and Diffusion

A key to successful pest-management programs is to design and support workable, cost-effective approaches to IPM. Research and especially extension programs in many countries have historically had a bias towards pesticide use. A review of existing public and private programs can help identify means for improving these programs, often with an eye on making them more farmer-participatory. To the extent that IPM programs hold promise for reducing environmental externalities, continuing public support can be justified as a means of equating social costs and benefits in pest-management programs. Policy analysis can help in designing more effective IPM research extension strategies.

Case Studies from the IPM CRSP

Case studies conducted on the IPM CRSP illustrate major types of IPM policy analysis. One case was completed in the Philippines, one in Jamaica, one in Ecuador, and one in Central America. The Philippines and Jamaica cases are briefly summarized below to illustrate a variety of approaches to IPM policy analysis. The Ecuador case, presented in Yamagiwa (1998), is similar to the Philippines case and is not summarized here. The Central America case, which explores the effects of technical barriers to trade in horticultural products, is still underway as this chapter is being prepared in 2004, but preliminary findings are discussed.

The Philippines: Analysis of direct and indirect pesticide subsidies and taxes

In the early 1990s, import tariffs were in place on all formulated pesticides and pesticide-active ingredients (called technical). It appeared that pesticides were being taxed, a policy that should have given an advantage to alternative IPM practices. However, the foreign exchange rate was also overvalued because the government was using exchange-rate controls and failing to devalue the peso sufficiently during times of inflation. Overvaluation had the effect of subsidizing imports, including pesticides, since all formulated and technical pesticides are imported in the Philippines. Therefore it wasn't clear whether the net effect of the tariffs and the exchange rate was to tax or subsidize pesticides, with attendant incentive or disincentive effects on IPM use. The IPM CRSP used procedures described in detail in Tjornhom et al. (1998) to assess the net effects of the policies.

In brief, an effective rate of protection (ERP) was calculated that included the effects of distorting policies for both tradable outputs (formulated pesticides in this case) and tradable inputs (technical pesticides in this case) by calculating the difference between value-added at the domestic price (including market distortions) and value-added at the border price (excluding market distortions) (see Appendix 1). The effective rate of protection depends on the tariffs for the technical and formulated pesticides and on the ratio of the border to the domestic price. The domestic price depends on the marketing margin, and therefore the effective exchange rate also depends on the proportion of the retail price that is accounted for by the product itself as opposed to transportation and processing costs.

The ERP can be calculated using a nominal exchange rate, a free-trade exchange rate, or a free-trade equilibrium exchange rate. The analysis in the Philippines study calculated ERPs using both the nominal and the free-trade equilibrium exchange rates. The latter is the exchange rate that would make the country's current account balance in the absence of tariffs, quotas, export taxes, and other trade restrictions for a given price of non-tradeables (See Appendix 1).

The effects of tariffs on formulated and technical pesticides are presented graphically in Figure 10-1. A tariff (t_f) on formulated pesticides (Figure 10-1a) increases the domestic price of formulated pesticides to PF¢ and reduces import demand and increases domestic supply for those pesticides. The effect is to increase the demand for technical pesticides (Q to Q¢ in Figure 10-1c) and for pesticide processing. A concurrent but smaller tariff (t_T) on technical pesticides increases the price of technical pesticides (to PT¢ in Figure 10-1c); however, the Philippines still does not produce any technical pesticides domestically. The effect of the technical pesticide tariff is to shift back the supply curve for formulated pesticides so as to reduce the demand for pesticide processing and technical (to Q^2) and increase the tariff revenue on formulated pesticides (by the dark shaded area in Figure 10-1a). It also generates tariff revenue on technical pesticides, and the price of pesticide processing is reduced to PP^2.

An overvalued exchange rate acts as an import subsidy. The effect on the formulated pesticide market can be illustrated graphically by reversing the effects of the import tariff on formulated pesticides illustrated in Figure 10-1. Pesticide price would be reduced, imports would increase, and pesticide processing and use of technical pesticides would decrease. The effects of this implicit subsidy on technical pesticides can also be illustrated by reversing the effects of the import tariff on technical pesticides illustrated in Figure 10-1. The price of technical pesticides would decrease and the supply curve for formulated pesticides would shift down to the right. The result

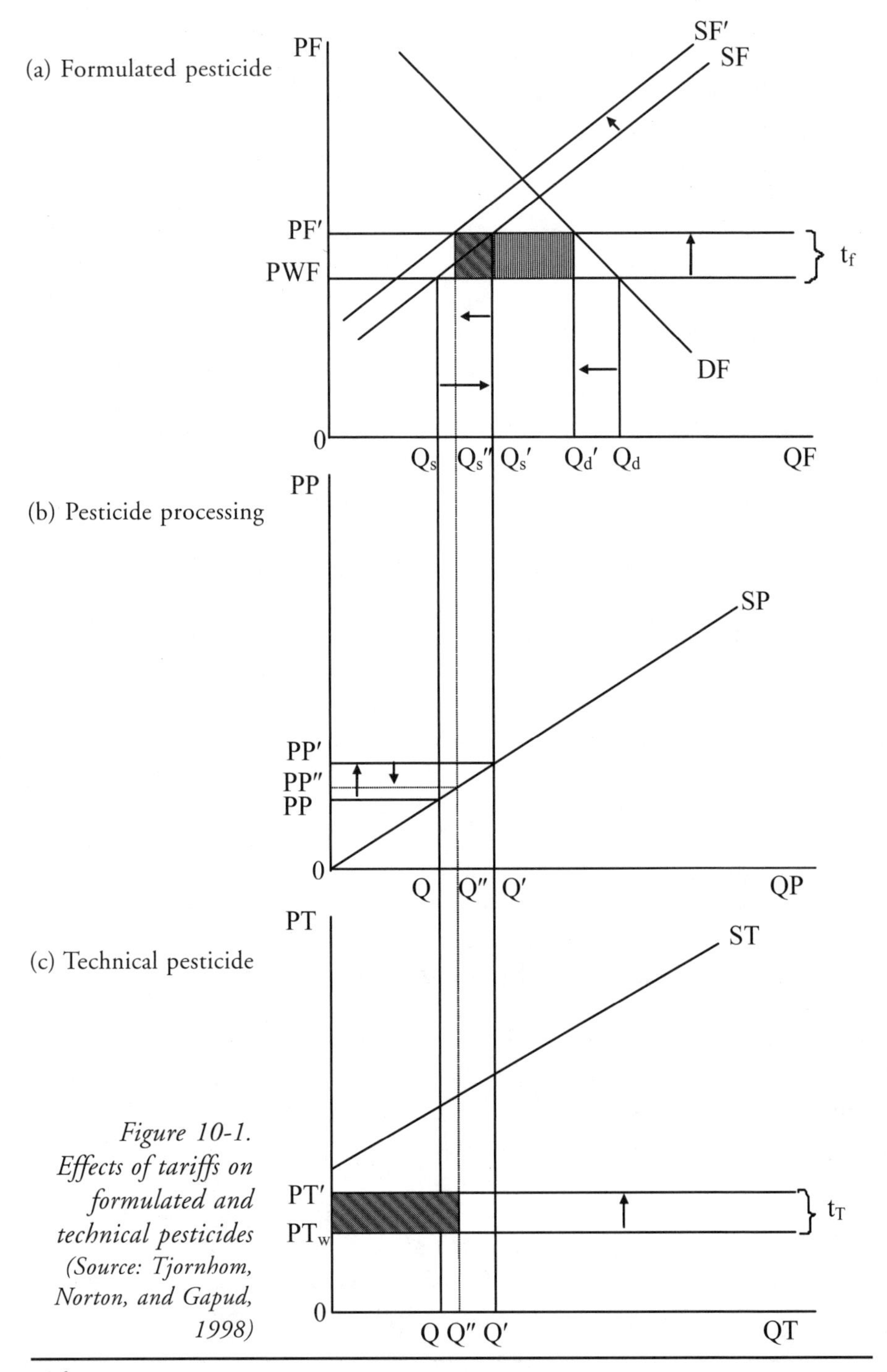

Figure 10-1. Effects of tariffs on formulated and technical pesticides (Source: Tjornhom, Norton, and Gapud, 1998)

would be an increase in domestic production of formulated pesticides and an increase in pesticide processing. However the implicit subsidy on technical pesticides does not affect the price of formulated pesticides.

Calculating welfare effects

The net benefits or costs and net retail price effects of market distortions in the pesticide market are illustrated in Figure 10-2. One can view Figure 10-1a as the wholesale level for formulated pesticides and the market represented in Figure 10-2 as the retail level lying above Figure 10-1a. Eight of the nine pesticides analyzed in the Philippine study are imported only as technical pesticides, and once formulated, these pesticides are not exported. The one pesticide, Cymbush®, that is imported in formulated form (and is not imported as technical) can also be analyzed at the retail level, as in

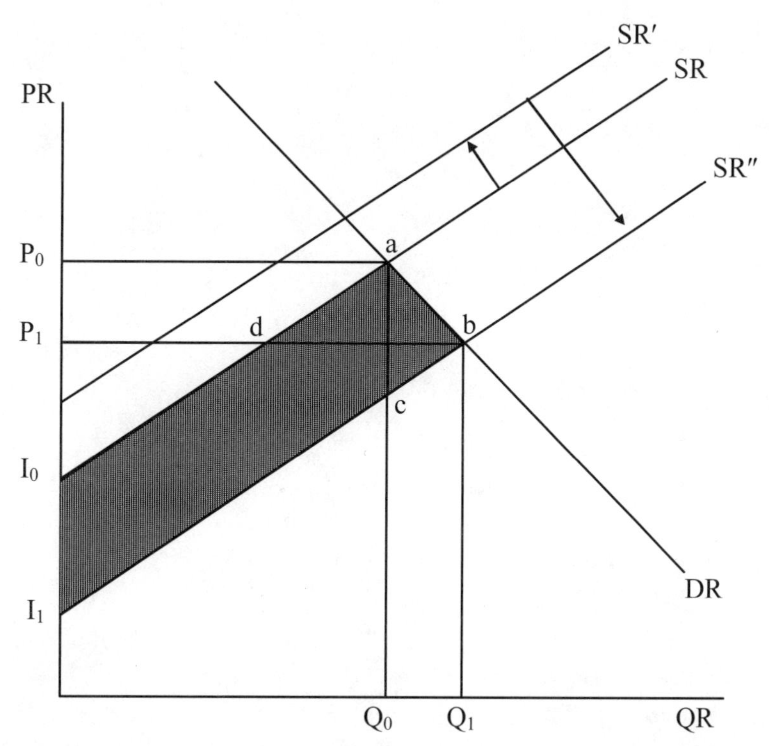

Figure 10-2. Economic Surplus (welfare) effects of the direct and indirect policy distortions in the pesticide market (Source: Tjornhom, Norton, and Gapud, 1998)

Figure 10-2, with the slope of the supply curve dependent on the nature of domestic marketing costs.

At the retail level, the tariffs on either formulated or technical pesticides shift the supply curve back from the free market equilibrium at P_0, Q_0, but the exchange rate overvaluation shifts it down to the right, resulting in a policy distorted equilibrium at P_1, Q_1. The change in producer and consumer surplus, assuming no environmental externalities, would be I_0abI_1, made up of the change in consumer surplus, P_0abP_1, and the change in producer surplus, $I_0dbI_1 - P_0adP_1$. In this case, "producer" surplus includes benefits to importers of technical pesticides, importers of formulated pesticides, producers of formulated pesticides using imported technical, pesticide distributors, and pesticide dealers. "Consumer" surplus includes benefits to individuals and companies that purchase pesticides at the retail level from pesticide dealers and individuals that purchase the crops on which pesticides are used.

The change in consumer, producer, and total economic surplus can be calculated as:

$$CCS = P_0Q_0Z(1+0.5Zn) \quad (6)$$

$$CPS = P_0Q_0(k-Z)(1+0.5Zn) \quad (7)$$

$$CTS = P_0Q_0k(1+0.5Zn) \quad (8)$$

where k is the vertical shift in the supply function as a proportion of the initial price, n is the absolute value of the price elasticity of demand, e is the supply elasticity, and $Z = ke/(e+n)$. Because the initial prices and quantities already reflect the influences of tariffs and exchange-rate policies, the vertical shift will be negative in the calculations and the surplus will reflect the benefits of removing the policies.

The effects of the policies on pesticide consumption and prices can be calculated and presented as well. The consumption of pesticides in the absence of intervention can be estimated as:

$$Q_0 = Q_1 / (1+Zn) \quad (9)$$

and the price that would have prevailed without the tariffs and exchange-rate overvaluation as:

$$P_0 = P_1 / (1-Z). \quad (10)$$

Data

Tariff rates for calculating ERPs were obtained from the Philippine Bureau of Agricultural Statistics (BAS). Border prices of formulated pesticides in dollars per liter and technical pesticides in dollars per kilogram were obtained from the Business Statistics Monitor's *Monthly Descriptive Arrivals Report.* Retail price of each of the nine pesticides was determined by surveying pesticide dealers in San Jose City in Nueva Ecija. To derive the price at which the pesticide producer sells the formulated pesticide to the distributor, the marketing margin between the producer and retail level (M) was subtracted from the retail price. Through interviews with the Director of the Fertilizer Pesticide Authority of the Philippines and representatives of pesticide companies in the Philippines, the marketing margin was estimated at 30 percent of the retail price of the formulated pesticide. The pesticide distributor buys the formulated pesticides at approximately a 30 percent savings off the suggested retail price and the distributor sells to the pesticide dealer at approximately a 15 percent reduction off the suggested retail price, but the total mark-up between the pesticide producer and the pesticide dealer is 30 percent. This figure was used as a rule of thumb to back off the retail price to estimate the producer's price of the formulated pesticide. Data for the equilibrium exchange rate calculations were obtained from the IMF International Financial Statistics yearbook (1995) and from Intal and Power (1991). Details on data collection and calculations are found in Tjornhom et al. (1998).

Results

The effective rate of protection under the actual market exchange rate and under the free-trade equilibrium exchange rate are summarized in tabular form in Tjornhom et al. (1998) by pesticide by year for a recent five-year period (1989–1993). The results indicate an average rate of disprotection for pesticide producers of 12 to 25 percent. In other words, pesticides were cheaper than they would have been without the combination of tariffs and exchange rate effects. The degree of divergence between the actual and the equilibrium exchange rate varied from –13.7 to –19.9 percent for 1984 to 1993 with a negative divergence indicating an overvalued exchange rate. The average amount of overvaluation was 17.7 percent. In the retail market, the vertical shift in the supply curve, taking into account both the overvaluation and the 5 percent tariff for technical pesticides, was estimated to be –13.6 percent. For the one imported formulated pesticide, it was –8.6 percent.

On average, retail price for pesticides was 6 percent lower and the quantity purchased 3.5 percent higher than they would have been without the distortions. These changes are based on an assumed pesticide supply elasticity of one and demand elasticity of –.5. When the demand elasticity is varied from –.25 to 1, quantity purchased varies from about +2 percent to +7 percent.

The estimated change in economic benefits to pesticide producers and consumers was positive for both producers and consumers if environmental externalities are not considered. The overvalued exchange rate more than offsets the tariff effects, resulting in a greater quantity consumed at a lower price, and the cost of producing the pesticide was reduced by more than the price reduction. Larger surplus values are associated with pesticides consumed in greater quantities, such as Thiodan, Azodrin, and Nuvacron. The demand elasticity had little effect on the total economic surplus, but did substantially affect the distribution of benefits between producers and consumers, with higher demand elasticities benefiting pesticide producers.

Conclusions

Results of the analysis indicated that pesticide producers received less for formulated pesticides than they would have if all ingredients had been manufactured in the Philippines. However, this tax on pesticide value-added did not appear excessive and was experienced by all pesticide producers. Therefore, tariff and exchange rate policies offered little deterrent to the importation of formulated and technical pesticides. In fact, the value of pesticide production was greater under the existing policy environment than it would have been in the absence of tariffs and exchange rate overvaluation. It should be noted, however, that the overvalued exchange rate put downward pressure on pesticide demand through its effects on reducing crop exports and increasing crop imports.

If the health and environmental costs associated with pesticide use were considered, they would have reduced the policy benefits accruing to consumers and producers and very well might have turned them negative (Cuyno, 1999). Two key policy conclusions emerged from the analysis. First, the relatively low level of net subsidies on pesticides in the Philippines (which likely would have been measured to be even lower if the analysis had included the exchange-rate effects on crop imports and exports) implied that such subsidies were providing little deterrent to the adoption of IPM in the Philippines. This was a significant conclusion given the emphasis in the Philippines on generating new IPM practices for vegetables. This study was undertaken in part because of a concern that, if significant pesticide subsidies exist, biological researchers developing IPM approaches would be wasting their time unless the policies were changed.

Second, given the expressed concerns over health and environment effects of pesticides in the Philippines, if a reduction in pesticide tariffs were to occur under international trade agreements, some other policy tools might need to be implemented to offset the resulting increased pesticide subsidy. Currently, several of the most toxic pesticides are banned in the Philippines (not the ones evaluated in this study). Enforcement of the ban has proven difficult, however, as is often the case with such regulations in developing countries. And, it is unlikely that a complete ban is desirable for the less toxic chemicals. The import tariff on pesticides has involved relatively low transaction costs, and it raises revenue for the government. One might argue that it makes sense as a "green tax" designed to raise the marginal private cost of pesticides up to their marginal social cost. If such a green tax is not allowed under the WTO, the government might alternatively impose a domestic sales tax on pesticides to serve the same purpose. Clearly, the exchange rate policy will be little influenced by pesticide issues; therefore the optimal policy tool is likely to be some sort of tax (tariff if allowed, because of the low transaction costs, or a domestic sales tax). Direct subsidies for IPM adoption are likely to be less advisable due to difficulties in defining IPM, government budget implications, and high transaction costs.

Jamaica: Analysis of Policies Affecting IPM Adoption

The Jamaican government is attempting to reduce pesticide use on farms by encouraging IPM adoption. Working with the Caribbean Agricultural Research and Development Institute (CARDI), with support from USAID and other sources, Jamaica has developed IPM systems for several vegetable and root crops, including hot pepper (*Capsicum chinense*), vegetable amaranth (*Amaranthus* sp.), and sweet potato (*Ipomoea batatas*).

In an IPM CRSP study, the net economic returns of adopting IPM for hot pepper, sweet potato, and amaranth were evaluated for small farmers in South Central Jamaica (Ebony Park, Clarendon, and Bushy Park, St. Catherine). The results provided a baseline for comparing effects of policy changes on the profitability of IPM adoption. Policy changes were simulated for a) a change in pre-clearance policy designed to lower health and safety trade barriers, b) the elimination of concessionary water rates to farmers,

c) lowering the Common External Tariff, and d) the elimination of a credit subsidy.

Reducing health and safety barriers to trade might be accomplished through farm level inspections. These inspections would increase crop labor requirements. The adoption of IPM can lead to higher labor use, and initiation of pre-clearance practices by farmers could compete with IPM practices for scarce labor, potentially reducing the profitability of IPM.

A reduction in the Common External Tariff might lower the cost of chemical inputs relative to labor, and lead to lower levels of IPM adoption. An elimination of the water subsidy would raise the price of water and could potentially affect the profitability of IPM because of irrigation required in the IPM system for sweet potato. A higher real interest rate resulting from the reduction of the credit subsidy could lower the cost of labor relative to chemical inputs because farmers typically borrow to purchase chemical inputs but not labor.

A mathematical programming model was used for the analysis. Representative farms were constructed for locations in Ebony Park, Clarendon and Bushy Park, St. Catherine. Information from the Ministry of Agriculture, and from interviews in 1998 with local farmers, IPM scientists, and government officials provided the data for the model. The effects of IPM on net returns were evaluated with and without policy changes.

Empirical Framework

The objective function of the programming model maximized the returns to fixed resources (owner labor and management, land and capital) above variable costs. Variable costs included the cost of seeds, fertilizers, pesticides, hired labor, and water. All costs and revenues within the models and the subsequent analysis were in 1997 Jamaican dollars.

The time horizon for the models was one year, and farm sizes were ten acres for the Ebony Park farm and five acres for the Bushy Park farm. The representative farm model for Ebony Park contained seven crops and ten production methods. It assumed that hot pepper, amaranth, and sweet potato could be grown by conventional methods or by using IPM. Corn, cassava, sugar cane, and pumpkin could be grown by conventional methods only. The Bushy Park farm model contained hot pepper and amaranth grown by conventional or IPM methods and corn, pumpkin, okra, and cucumber grown by conventional methods.

The farmers in Ebony Park and Bushy Park borrowed money from family members or from the Agricultural Credit Bank (ACB) system to purchase inputs such as seeds and chemicals. The subsidized interest rate was set at 3%, and the unsubsidized rate was set at 25.5%.

Resource, technical, and risk constraints were included in the model. The resource constraints limited the farmer's access to land, owner labor, and water. Water and labor constraints were specified for each month. The technical constraints forced the farmer to use the prescribed inputs and production practices to produce and sell the output. The risk constraints limited the total acreage for each crop based on farmers' concerns about risk. All farmers voluntarily constrained the total acreage of land for each crop because of market uncertainty.

IPM systems

The IPM package for sweet potato, which was focused on the sweet potato weevil, *Cylas formicarius*, includes cultural, and biological components such as pheromone traps, removal of alternate hosts, harvest timing, and field sanitation. (See Ogrodowczyk et al. for details). Marketable yield increases of 15% were assumed for the IPM systems, based on experimental results of Lawrence, Bohac, and Fleischer (1997).

The IPM package for hot pepper includes cultural, mechanical, biological, and chemical components to help manage Tobacco Etch virus, Potato Y virus, mites (*Polyphagotarsonemus latus*), and gall midges (*Contarinia lycopersici* and *Prodiplosis longifila*). The hot pepper IPM system was estimated to lengthen the harvest season by an additional month and increase yield by 30% (Martin, 1998). The IPM system for vegetable amaranth included chemical, mechanical, cultural, and biological components (Clarke-Harris et al., 1997). Yields were estimated to increase by 30%.

Policy scenarios

The models initially assumed existing policies, and then four changes were evaluated: pre-clearance practices by farmers, elimination of the $JM11.39 per hour subsidy on water sold to farmers, lowering of the Common External Tariff by 11.25%, and elimination of the 22.5% credit subsidy.

Pre-clearance

The USDA Animal and Plant Health Inspection Service (APHIS) requires that all agricultural imports to the United States pass phytosanitary inspections for insects and diseases. Exporting countries can choose to build a domestic facility, called a pre-clearance station, to allow agricultural produce to be inspected before it is exported to the United States. Produce passing through the facility directly enters the U.S. market. The Jamaican government requires all agricultural exports to pass through a pre-clearance station in Kingston or Montego Bay. Farm-level inspections are designed to work in conjunction with the pre-clearance facilities. Pre-clearance affects the labor required for production, price received for goods, and quantity exported. If produce is found to be infected/infested at pre-clearance, the farm is automatically inspected on all future shipments from the farm. To prevent such inspections, farmers check their produce before shipping it to the pre-clearance station. Labor requirements during harvest season were increased by 25% to simulate farmers' self inspection. Farmers inspecting their harvest received a higher export price (2.5% higher), reflecting the increased confidence of the exporters purchasing from the farmer (Reid, 1998). Marketable yield is raised by about 10%.

Common External Tariff and the Uruguay Round Agreements

Goods imported into Jamaica from non-CARICOM countries are charged the Common External Tariff (CET). In June of 1998, the CET was 80% of the value of the goods imported. During the Uruguay Round negotiations of the General Agreement on Tariffs and Trade (GATT), Jamaica, along with the other CARICOM nations, made concessions limiting the CET and agreed to reduce it by 20% over seven years. According to government officials, the annual reductions had not been specified but a 11.25% cut was feasible. In the empirical model, the CET was lowered by 11.25%, leading to a 5% reduction in the cost of pesticides, a 5% fall in the price of fertilizers, and a 4.75% decrease in the cost of machinery services.

Water Subsidy

The National Irrigation Commission (NIC) rates for agricultural water use include a service charge and a demand charge. From September 1996 until August 1998, the service charge was $JM12.11, $JM24.23, and $JM30.29 per month for farms less than five acres, between five and ten acres, and more than ten acres, respectively. The representative farms purchase water from the NIC for $JM8 per hour including service and pumping costs. The elimination of the water subsidy is accomplished by restricting the farmers in both models to pump their water from nearby river sources at a cost of $JM19.39 per hour.

Credit Subsidy

The Agricultural Credit Bank provides subsidized financial assistance for farmers. Loans to farmers are categorized between small farmers and large farmers. Both representative farms were classified as small farms and were limited to $JM 100,000 of subsidized credit at the real interest rate of 3%. When the credit subsidy was eliminated, farmers paid a real interest rate of 25.5%.

Results

Without IPM, the Ebony Park farm (Scenario One) earned gross revenues of $JM565,074 and returns to fixed resources of $JM452,995. The Bushy Park farm earned gross revenues of $JM 710,607 and returns above variable cost of $JM 643,863. Ogrodowczyk et al. provide details on quantities produced of each crop, amount borrowed, water use each month. Neither land nor subsidized credit was a binding constraint.

With IPM, the Ebony Park farm (Scenario Three) earned gross revenues of $JM 613,725 and returns to fixed resources of $JM497,808. The farmer increased borrowed capital from $JM 58,224 to $JM 60,492 at the subsidized interest rate but did not reach the limit. The farmer increased the total water usage, and total labor use rose by 6.5%. With IPM, the Bushy Park farm (Scenario Four) earned gross revenue of $JM 755,960 and returns above variable cost of $JM 696,009. Land was a binding constraint and an additional acre would be used entirely for IPM amaranth production. In contrast to the Ebony Park farm, the Bushy Park farm decreased borrowed capital by 3.7%.

To examine the robustness of the results from IPM adoption, sensitivity analysis was conducted for changes in production and profits from altering chemical requirements, yields, labor, and storage requirements. Chemical reductions resulting from IPM were eliminated, increases in IPM yield were lowered, and increases in IPM labor requirements were raised. The analysis indicated that even with large unfavorable changes in chemical use, labor requirements and yield increases, IPM continued to be more profitable than conventional systems.

Parameters in the model were altered to simulate the effects of policy changes. In the pre-clearance policy scenario, three groups of parameters were altered: harvest labor (increased by 25%), export prices (increased by 2.5%), and export quantities (increased by 10%). With pre-clearance but without IPM, the Ebony Park farmer earned returns by 10% compared to returns without pre-clearance; in Bushy Park, 2%. When the pre-clearance policy was in place and IPM was available, the Ebony Park farmer increased returns to fixed assets by 21% compared to no pre-clearance and no IPM; the Bushy Park farmer, 10%. IPM adoption had similar effects on input use with or without the pre-clearance policy. Both farms had modest increases in water and labor use when IPM was adopted. Bushy Park had a modest reduction in borrowed capital while Ebony Park had a slight increase in borrowed credit.

Several changes in the price of chemicals, net returns, and borrowed capital occurred as the Common External Tariff (CET) was lowered. However, the farms in both areas adopted IPM at the same levels as with the original CET. The lower cost of chemicals did not become a disincentive for IPM adoption.

The elimination of the water subsidy was hypothesized to impact the profitability of IPM because irrigation is a component of the IPM package for sweet potato and because corn produced for the IPM hot pepper system requires irrigation. Without the water subsidy and with access to IPM, both farms adopted IPM at the same level as was adopted under the baseline with a water subsidy. Higher water costs did not affect IPM adoption. Likewise, when the credit subsidy was eliminated, both farms adopted IPM at the same level as under the baseline with a credit subsidy. Input use changes with IPM adoption were the same without the credit subsidy as with the subsidy.

In summary, the results showed that the IPM systems were more profitable than conventional production practices. This result was not sensitive to changes in the yield increase from IPM, the increased IPM labor requirement, the reduced chemical requirement of IPM, or crop storage requirements. The economic liberalization reforms now being considered by the Jamaican government would have little effect on the profitability of IPM adoption except for pre-clearance, which enhanced the profitability of IPM. Reductions in the CET or removal of the credit or water subsidies lowered the profitability of IPM adoption by small amounts but did not affect the optimal level of IPM adoption. The policies of eliminating the water subsidy, lowering the CET or eliminating the credit subsidy would not require offsetting policies for the adoption of IPM by Jamaican farmers. However, extension efforts aimed at teaching farmers about the benefits of pre-clearance could be coordinated with methods for encouraging the incorporation of the IPM technologies to increase the rate of IPM adoption.

Central America: Technical Regulations in Agricultural Trade

Technical barriers to trade (TBTs), which include sanitary and phyto-sanitary (SPS) regulations as a subset, are regulations that can be disguised measures of protection. The Agreement on the Application of Sanitary and Phyto-sanitary Measures (SPS Agreement) and the Agreement on Technical Barriers to Trade (TBT Agreement) were included in the WTO Agreements in 1994 to help sort out the legitimate use of technical regulations from the practice of economic protection of domestic producers. The TBT Agreement defines technical regulations underlying TBTs to be regulations that describe product characteristics or their related processes and production methods (WTO, 1994a). SPS measures are measures applied to products entering into a country with the aim of protecting human, animal, and plant life from diseases or contaminated foods, beverages, feedstuffs, animals, or plants (WTO, 1994b). These agreements have influenced agricultural and food trade for developing countries and hence incentives to adopt IPM.

In addition to the restriction that risk assessment must be based on science, the SPS Agreement promotes several concepts that aim to facilitate trade expansion. (Yamagiwa, 2004).

A two-part analysis of Central American SPS issues was conducted on the IPM CRSP. One part was a review of data on export rejections due to SPS issues and complaints against specific countries. The second was a set of interviews (approximately 100) with representatives of domestic and international institutions, NGO representatives, and industry people in the region to obtain their perceptions and concerns on SPS and other trade issues. Interviewees provided opinions on the importance of technical regulations in trade and the impact of international institutions and the Central American Free Trade Area on trade-related technical regulation issues.

A large amount of data was summarized and reported in Yamagiwa. For example, U.S. refusals of Central American imports by country and reason were collected and are reported in Table 10-2. Guatemala stands out as a country with many refusals due to pesticide residues, implying a need for IPM and for a pre-clearance program. The interviews led to several conclusions. First, international institutions have positively influenced the degree to which Central America has been able to manage trade-related technical regulations, but additional assistance is needed. Second, regional

Table 10-2. U.S. Food and Drug Administration (FDA) import refusals by reason

Category of reason for refusal	Country of origin					
	Costa Rica	El Salvador	Guatemala	Honduras	Nicaragua	Total
Labeling	4	28 (12)	69 (37)	23 (18)	12 (7)	136(78)
Pesticide residues	24	0	64	8	0	96
Colorant	0	8	1	0	0	9
Poisonous	3	1	0	2	2	8
Filthy/insanitary	13	13	6	15	10	57
Aflatoxin	0	0	1	0	0	1
Microbiological cont'n	2	10	5	25	8	50
Administrative req't	12	12	21	5	2	52
Total	58	72 (56)	167 (135)	78 (73)	34 (29)	409(351)

Notes: Data for the period November 2002 through October 2003. The total number of refusal reason categories is greater than the number of refusals as one refusal may be due to multiple refusal reason categories. Numbers in parentheses refer to the citing of distinct category of reasons for a given refusal (i.e., given a refusal, a category is counted only once even when multiple reasons within a category are cited).

Source: Yamagiwa (2004) elaboration based on FDA 2003 data.

institutions seem also to have reduced the negative impact of technical regulations on agricultural trade. Third, the higher incidence in the past of complaints by regional trade partners against themselves than of complaints by the United States against Central American countries may reflect the ability of the United States to utilize multilateral institutions to try to resolve disputes with the region. The recent reduction in intraregional complaints reflects an increased capacity of regional institutions to also serve as efficient communication channels.

Interviewees expressed the view that technical regulations are important in Central American agricultural trade, especially for non-traditional agricultural export crops. The high incidence of U.S. SPS measures combined with low U.S. tariffs on Central American imports suggests that they are more important than direct protection measures in influencing trade flows. In intraregional trade disputes, almost half of all disputes result from technical regulations. In summary, policy analysis that includes rather straightforward data analysis combined with a set of interviews can be used to help identify the relative importance of technical barriers to trade as compared to other barriers and can help guide policy changes to facilitate trade and economic incentives to adopt IPM.

Conclusions

The cases presented illustrate a few of a wide diversity of policy analyses that could be undertaken with respect to pest management. They are among the first types of policy analyses to consider because each, using different methods, focuses on the issue of policy effects on IPM adoption. If policies create barriers to IPM adoption, such that there is little economic incentive to adopt, there may be little return to IPM research and extension. Among the various types of policies, regulatory mechanisms are critical to IPM and must be examined country by country.

The policy issue of defining the appropriate mechanism for diffusing results of IPM research is also of vital importance and has been little studied. Different approaches have been advocated such as village extension agents, training and visit systems, and farmers' field schools. However, it difficult to find a careful comparative assessment of the appropriate roles of different approaches in relation to the types of technologies involved and other factors. Policy analysis with respect to IPM extension is in its infancy.

References

Agne, Stefan, G. Fleisher, F. Jungbluth, and H. Waibel. 1995. Guidelines for pesticides policy studies: A framework for analyzing economic and political factors of pesticides use in developing countries. *Pesticide Policy Project Publication Series, No 1.* Hannover, Germany: Institute of Horticulture Economics, February.

Bureau of Agricultural Statistics (BAS). 1995. Report on the Performance of Agriculture, January–December 1994. Department of Agriculture, Manila.

Business Statistics Monitor, 1989 to 1995 issues. Monthly Descriptive Arrivals Report: Insecticides, Fungicides, Herbicides, Disinfectants, etc. Manila, Philippines.

Center for Research and Communication (CRC), Agribusiness Unit. 1994. Philippine Agribusiness Factbook and Directory 1993–1994. Southeast Asian Science Foundation, Inc.

Clarke-Harris, Dionne, Janice Reid, and S. Fleischer. 1997. IPM systems development: Callaloo, *Amaranthus* sp. *IPM CRSP: Fourth Annual Report 1996–1997.* Blacksburg, Virginia: Virginia Tech, pp. 167-182.

Corden, M. 1971. *Theory of Protection*, Chapter 3. Oxford: Oxford University Press.

Cuyno, Leah, C.M. 1999. An economic evaluation of the health and environmental benefits of the IPM CRSP program in the Philippines. Ph.D. Thesis, Virginia Tech.

FAO (Food and Agriculture Organization). 1990. *International Code of Conduct on the Distribution and Use of Pesticides*. Amended version, Rome.

Intal, P.S., and J.H. Power. 1991. The Philippines in *The Political Economy of Agricultural Pricing Policy: Asia. Volume 2,* ed. A.O. Krueger, M. Schiff, and A. Valdes. Washington, D.C.: World Bank.

International Monetary Fund. 1995. *IMF Financial Statistics Yearbook*. Washington, D.C.

Lawrence, Douglas. 1998. United States Agency for International Development, Animal and Plant Health Inspection Service. Kingston, Jamaica. Interview, 29 June.

Lawrence, Janet, J. Bohac, and S. Fleischer. 1997. Sweet potato weevil Integrated Pest Management, In *Progress in IPM CRSP Research: Proceedings of the Second IPM CRSP Symposium*. Guatemala City, Guatemala: Integrated Pest Management – Collaborative Research Support Program, May, pp. 37-45.

Lawrence, Velva. 1985. Organochlorine residues in Rio Cobre, Jamaica and the effect of Dieldrin on the physiology of shrimps. M.S. Thesis. Kingston, Jamaica: University of the West Indies, 76 pp.

Martin, Raymond. CARDI, Kingston, Jamaica. Interview, 17 May 1998.

Ministry of Agriculture. Draft for the *World Trade Organization Trade Policy Review*, Kingston, Jamaica: Ministry of Agriculture.

Ogrodowczyk, Joseph, Darrel Bosch, Janet Lawrence, and Dionne Clarke-Harris. 2000. Policies affecting adoption of Integrated Pest Management by Jamaican farmers. Department of Agricultural and Applied Economics, Virginia Tech, Blacksburg, Virginia, July. 31 pp.

Reid, Charles. 1998. Jamaican Exporters Association, Jamaica. Interview 30 June.

Shillhorn van Veen, T.W., Douglas A. Forno, Steen Joffe, Dina L. Umali-Deininger, and Sanjiva Cooke. 1997. Integrated Pest Management: Strategies and policies for effective implementation. ESDS Series No. 13. Washington, D.C.: World Bank.

Tjornhom, Jessica, George Norton, and Victor Gapud. 1998. Impacts of price and exchange rate policies on pesticide use in the Philippines, *Agricultural Economics* 18:167-175.

World Trade Organization (WTO). 1994a. The WTO agreement of technical barriers to trade (TBT Agreement). <http://www.wto.org/english_docs_e/legal_e/17-tbt_e.htm>

World Trade Organization (WTO). 1994b. The WTO agreement on the application of sanitary and phytosanitary measures (SPS Agreement). <http://www.wto.org/english/tratop_e/sps_e/spsagr_e.htm>

Yamagiwa, Takayoshi. 1998. Analysis of policies affecting pesticide use in Ecuador. MS Thesis, Virginia Tech.

Yamagiwa, Takayoshi. 2004. A systematic assessment of the incidence of technical regulations in Central American agricultural and food trade. Blacksburg, Va.: Department of Agricultural and Applied Economics, Virginia Tech.

— 11 —
The Role of Institutionalized Pre-Shipment Inspection Programs
in Achieving Sustainability in Non-Traditional Agricultural Export Markets

Glenn H. Sullivan, James Julian, Guillermo E. Sánchez,
Steven Weller, and George W. Norton

Introduction

Production and export of horticultural products have increased significantly in developing countries over the past three decades. Fruit and vegetable production in developing countries increased from 256 to 960 million MTs from 1970 to 2002, while exports jumped from 1.9 to 6.5 million MTs (FAOSTAT database). Demand for these high-value commodities has been stimulated by income growth, reductions in transportation costs, and, in some cases, increased market access. Export production can generate foreign exchange, increase producer incomes, and provide employment for rural poor. Importing countries can benefit from increased supplies of products that otherwise might be expensive in the off-season.

Rapid growth in horticultural products has been accompanied by heavy use of pesticides, and by concerns over health effects associated with pesticide use and abuse. Heavy pesticide use occurs, in part, because pests are numerous on horticultural crops, often attacking the fruit itself and reducing market value and yield of a high-value commodity. The safety of agricultural workers who apply pesticides may be compromised, particularly in flower production under greenhouse conditions. Potential food safety risks from pesticide residues are a concern for importers of fresh fruits and vegetables, and a market-risk factor for exporters who may have shipments rejected if residues exceed allowable limits.

Countries face a difficult task in trying to minimize pesticide residues and maintain product quality, while eliminating the presence of pests in horticultural export shipments. Interceptions of exotic or unidentified pests in fresh produce imports can result in the country being quarantined for that commodity, with all shipments requiring fumigation prior to entry into the United States. Repeated violations of allowable chemical residue levels can result in the automatic detention of shipments from a country until it can prove convincingly that corrective measures have been implemented. Developing countries are especially vulnerable to detentions because official programs designed to prevent the export of harmful organisms and/or food products may not be routine for fresh non-traditional agricultural exports (NTAEs). Therefore, these countries must set up exclusionary procedures to prevent horticultural pests or harmful chemical residue levels from contaminating agricultural exports.

The tight scrutiny of horticultural imports into the United States market is the result of increased food safety and agricultural quarantine concerns that have led to strengthened U.S. sanitary and phyto-sanitary (SPS) regulations in an attempt to insure the safety of the food supply (U.S. General Accounting Office, 1998; Institute of Medicine National Research Council, 1998). Recent events have also catalyzed the approval of new laws directed to ensure the safety of the U.S. food supply. The objective of The Public Health Security and Bioterrorism Preparedness and Response Act of 2002 (Bioterrorism Act) is to provide the Food and Drug Administration (FDA) with a detailed traceback capability in the event of a terrorist attack to the U.S. food supply or food-borne illness outbreaks. These regulations protect domestic consumers and domestic industry interests from hazardous contaminations that impact human health and from exotic pests that pose serious threats to the importing country's agricultural sector or environment. Also, in the current trade environment, it is difficult for countries to arbitrarily implement tariffs or quotas directly to restrict market access in an effort to influence competition. Consequently, SPS safety standards may occasionally be implemented as non-economic barriers to restrict market access as well (McDowell and Martinez, 1994). Even when imposed for the right reasons, SPS standards can constrain trade, and developing country exporters and producers must meet these standards in a science-based and market-driven manner if NTAE production is to grow sustainably and competitively in the global environment.

The FDA's Hazard Analysis and Critical Control Point (HACCP), and the APHIS (U.S. Department of Agriculture's Animal and Plant Health

Inspection Service) Pre-clearance programs represent science-based risk-management approaches to safe food production. HACCP programs are site-specific plans through which hazards to the health of consumers are prevented by applying science-based control mechanisms along the entire production system, from production of raw materials to a finished product ready for consumption. "Critical control points" are those specific points along the food production, manufacturing, and transportation system where hazards can be controlled or minimized through appropriate control measures. Pre-clearance protocols are developed along commodity-specific lines; their purpose is to provide point-of-origin (pre-shipment) inspection, including the entire system from production through post-harvest handling and export. IPM protocols have been developed recently for several horticultural commodities (such as mangoes) in developing countries to fit into these pre-clearance systems in order to satisfy the need for minimizing residues and providing the necessary pre-shipment documentation. These systems have required cooperation between the public and private sectors and between exporters and importers. Two examples from the Integrated Pest Management Collaborative Research Support Program (IPM CRSP) are provided below to show how applied research can reduce pesticide use, pesticide residues, and export barriers. The first example involves snow-pea production and exports from Guatemala, while the the second addresses hot-pepper production and exports from Jamaica.

Snowpeas in Guatemala

Commercial production of non-traditional fruits and vegetables for export has been the fastest growing segment of agriculture in Central America for the past 20 years. NTAE production has grown at an annual rate of approximately 15 percent and now totals roughly U.S.$ 400 million annually. Costa Rica, Guatemala, and Honduras have been the main regional participants in the commercial expansion of this production. This expansion has played an increasingly important role in the region's economic development, providing increased employment and income throughout the rural and small farm sectors. However, horticultural exports from Guatemala in particular have been plagued by detentions and rejections at U.S. ports, due both to excessive or unallowable pesticide residues and to the presence of pests in the food shipments. These detentions have especially been a problem for fresh vegetables; while Guatemalan fruit exports have maintained or increased their U.S. market share since the early 1990s, the country's fresh vegetable U.S. market share has exhibited a constant rate of decline (Julian et al., 2000).

The fruit and vegetable segments of the NTAE market represent two different approaches to the production and marketing of NTAE crops. The fresh-fruit sector has implemented programs to monitor and control sanitary and phyto-sanitary problems before the products are exported (Murray and Hoppin, 1992; Programa de Proteccion Agricola y Ambiental, 1995), such as the mango pre-clearance program in Guatemala. The fresh-vegetable sector has taken a more casual approach to SPS concerns, thus leaving their economic fate in the hands of the importing country's inspection personnel and regulatory policies. Consequently, fresh-vegetable export volumes from Guatemala have not grown at the pace of its regional competitors, such as Costa Rica, who have seen exports grow rapidly. Other competitors, such as the Dominican Republic, once plagued by import detentions (Murray and Hoppin, 1992), and Peru have increased fresh-vegetable exports to the United States as well (Julian et al., 2000).

Snowpeas (*Pisum sativum*), a primary Guatemalan vegetable export, have been under automatic detention by the FDA since 1992, mainly caused by excessive or unlabeled pesticide contamination. In addition, from 1995 to 1997 Guatemalan snowpea imports were quarantined (rejected) by the U.S. Department of Agriculture (USDA) when the presence of leafminer larvae (*Liriomyza huidobrensis* B.) was discovered in snowpea shipments. Guatemala lost US$ 35 million per year during the ban.

In the midst of the crisis, in 1996, the USAID IPM CRSP, the Government of Guatemala, and the Foreign Agricultural Service-USDA formed a strategic alliance to provide research and technical assistance to resolve the snowpea leafminer quarantine problem. Research protocols were established, demonstrating that the leafminer species discovered during the inspection was not exotic to the United States and consequently did not constitute a threat to U.S. agroecosystems. Since many secondary pests (such as *L. huidobrensis* in Guatemala snowpea fields) become primary pests due to excessive pesticide use, IPM research was undertaken and strategies developed to reduce pesticide use and residues on snow peas and to enhance product quality. As a result of this research, field-production and post-harvest management guidelines were designed and integrated into a "Pre-inspection" program. Pre-inspection refers to a pre-shipment quality-control program through which production, handling, and export of a given product is supervised by local authorities. A pre-inspection program is typically a voluntary, pro-active approach, certified by a recognized authority in the country-of-origin, that can lead to the establishment of a pre-clearance program. Currently over 40 percent of all snowpea production in Guatemala is accomplished under IPM CRSP-developed guidelines.

Snowpea IPM systems in Guatemala have been included in integrated crop-management (ICM) demonstration and training programs, with practices such as pest identification and monitoring, trap cropping, mechanical and physical pest-management tactics, soil disinfestation, use of bio-rational pesticides, and germplasm selection.

In Guatemala, about half the snowpea production is provided by one of three systems: (1) farms that both grow and ship; (2) cooperatives that market for many producers, and (3) growers who produce under contract to export firms. These supply channels generally produce snow peas with good pre-inspection protocols. The other half is supplied by independent producers who sell their product in an open market after harvest. These producers see no need to adopt any quality-control procedures, since any traceback possibility is lost once their product is mixed, by the intermediary, with that of many other independent growers. This is the main reason that many of the Guatemalan snow peas are automatically inspected at U.S. ports-of-entry. The IPM program, the Government of Guatemala, and private

Figure 11-1. Packing snowpeas for export in Guatemala

exporters are working together to improve practices in this open-market group. For those growers who have been trained in IPM and pre-inspection protocols, rejections at U.S. ports have been reduced by 50 to 75 percent, reflecting reduced pests and pesticide residues.

Production Issues Impacting Market Competitiveness

In assessing the baseline issues affecting competitiveness of Guatemalan snowpeas, the IPM CRSP found that heavy reliance on pesticides in pest management was a primary factor. For example, Guatemalan fresh vegetable shipments to the United States historically have had a high rate of FDA detentions due to excessive pesticide residues, often from pesticides not registered for use on those crops (Murray and Hoppin, 1992; Thrupp et al., 1995. Between 1984 and 1994, FDA detained 3,168 shipments of fruits and vegetables from Guatemala because of potential residue violations. The estimated value of these shipments was $17,972,000 (Thrupp et al., 1995). In 1996, FDA detained 641 Central American fruit and vegetable shipments; 575 of these were from Guatemala. This amount is significantly more than the 155 Central American shipments detained at U.S. ports-of-entry in 1990, 136 of which were from Guatemala. It was during this period that the leafminer outbreak intensified in the snowpea sector of Guatemala. Findings during the baseline assessment also showed that the high number of FDA detentions related to pesticide violations experienced by Guatemalan exporters was related to the absence of both effective integrated IPM/ICM protocols and adequate technology/information transfer to the producer level. The concurrent decline of competitiveness in Guatemala snowpea exports to U.S. markets indicated the importance of achieving effective technology transfer and the generation of production guidelines to ensure that producer practices are consistent with current market expectations. A holistic, total systems approach was needed to assure maximum performance.

The IPM CRSP in Central America, in collaboration with the Government of Guatemala/Ministry of Agriculture, Agricultural Research Fund of the Guild of Exporters of Non-Traditional Products (ARF/AGEXPRONT), the Universidad del Valle de Guatemala, and FAS-USDA, developed integrated pest/crop management (IPM and ICM) guidelines that would enable producers and exporters to comply with current SPS regulations in U.S. markets.

This inter-institutional effort was to also pursue the development of long-term solutions to insect-control problems that rely less on chemical control methods and subsequently result in greater sustainability at all levels within the NTAE sector. To achieve these goals, nine field-test sites (each 1,100 m^2) were established in the Departments of Chimaltenango and Sacatepequez in the Guatemalan Highlands, which account for nearly 80 percent of all snowpea production in Guatemala. In all cases, the most commonly utilized pest-control strategy had been to rely on the application of chemical pesticides, using a 7- to 10-day calendar-programmed schedule, with up to 14 pesticide applications per crop cycle. Most participating producers were small family farming operations with less than 0.5 hectare of snowpea production. Only one participating producer was a larger commercial operation with nearly four hectares of snowpea production. Few were acquainted with IPM strategies, and most relied heavily on agrochemical distributors for their pest-management information.

The most prevalent insect pressure was from leaf miners (*L. huidobrensis*), with some presence of thrips (*Frankliniella occidentalis, Thrips tabaci*). Ascochyta (*Ascochyta pisi*) leaf and pod blight were the most common fungal diseases. The level of insect and disease control was similar in both the IPM and control case-study test plots. However, pesticide applications in the IPM CRSP plots were significantly reduced, averaging about one-third the number of applications in the control plots (Sullivan et al., 1999, 2000). The IPM CRSP plots required an average of only 3.7 pesticide applications to fully achieve insect-pest management objectives, while the traditional chemical control plots required an average of 10.4 pesticide applications to achieve the same objectives. This chemical application reduction resulted in lower production costs and increased returns to household labor for the producers. In addition, snowpea yields were 23.4 percent higher in the IPM plots on average compared to the control plots. Moreover, the product quality was found to be higher in the IPM plots as measured by marketable yields at the shipping-point grading facilities. Product rejections at the shipping point averaged 6 percent less from the IPM plots. Snowpeas were not the only crop addressed. IPM CRSP research also addressed the whitefly-geminivirus complex that had crippled the tomato industry in some areas of eastern Guatemala.

Results from these commercial-sized field tests confirm the underlying premise of IPM, that immediate economic gain and long-term sustainability are reinforced through proper integration and management of existing production technologies in a holistic system. However, the adoption of this more effective IPM-based approach to crop management had been voluntary, and consequently implementation by producers has not reached all growers, as discussed above. In light of this low adoption rate, further research was conducted to assess the economic impact of establishing uniform production and post-harvest handling protocols, and the institutionalization of formalized CPP and HACCP programs for snow peas.

Pre-Shipment Inspection (Pre-inspection) Program

The Pre-shipment inspection program (or pre-inspection) was designed for snowpeas in Guatemala, in accordance with phytosanitary guidelines protocols. These protocols, for trace-back purposes, initially assign a code to each grower or supplier under contract with the packing facility/exporter. Field guidelines dictate that farmers follow a set of field-production guidelines, under the supervision of trained field inspectors. If farmers meet the minimum field-management requirements, including unexpected pesticide residue testing, a field-of-origin certificate is issued that enables producers to deliver snowpeas to the packing facility. At the packing facility, all handling and grading meets specified market and sanitary standards, with phyto-sanitary inspection following APHIS-approved sampling procedures. An area in the packing plant is set aside for inspection and, following classification and packing, a sample of the export product is tested for the presence of unwanted biotic agents (hitch-hikers). Rejected snowpeas are classified as to the cause of rejection, and packing plants maintain records of grading and phyto-sanitary results of inspections. Once the export product passes all pre-inspection criteria, each individual box (final packing container) is stamped with a seal of approval, indicating that their contents have been inspected in the field and during packing and meet sanitary and phyto-sanitary norms.

The pre-inspection protocols call for inspectors to be paid through private funds, but supervised by the proper local regulatory agency. In Guatemala, this agency is the Integrated Program for the Protection of the Environment and Agriculture-PIPAA-, a joint private sector/Ministry of Agriculture entity. PIPPA is the officially appointed entity to handle pre-inspection program implementation, compliance, and enforcement in non-traditional agricultural export crops.

Once PIPAA stamps its seal of approval, the product ready to be shipped is stored in a "clean" cold room (see Figure 11-2, page xxii) separate from the "dirty" cold room where un-graded and un-packed products are stored. Packing plants must meet hygiene and quality standards established through HACCP evaluations. They must keep records of all exported product, including domestic inspection results as well as inspection and interception reports from the port of entry.

In summary, the Pre-Shipment Inspection program is an integration of holistic, total systems performance protocols, grower training, technology transfer, and policy development and enforcement procedures. These items are combined to implement and enforce minimum quality and safety standards for snowpeas prior to their shipment from Guatemala.

Jamaica Pre-Shipment Inspection for Hot Peppers

The Caribbean region, including Jamaica, is exporting increased quantities of vegetables, including hot peppers (*Capsicum* spp). Because Jamaican peppers have arrived at U.S. ports infested with gall midge (*Contarinia lycopersici* and *Prodiplosis longifila)* (more than 100 shipments were intercepted with the pest in 1998), mandatory fumigation was instituted by the U.S. Animal and Plant Health Inspection Service (APHIS). Pepper exports from Jamaica declined by more than two-thirds from 1997 to 2000 as a result of the added cost of this fumigation. In response, the USAID-funded IPM program and several agencies of the Jamaican government have developed a multi-faceted IPM program, including a pre-clearance program in Jamaica. As a result, shipments found to be infested with gall midge dropped by more than 90 percent, and the mandatory fumigation was removed by APHIS in 2002 under several conditions. These conditions include, among others, a requirement that only growers participating in the field-control program would be allowed to ship without fumigation and growers with shipments rejected for the midge would be removed from the program. IPM strategies that involve: (1) improving cultural practices and reducing pesticide use in the field, (2) substituting a less costly and environmentally safe fumigant than methyl bromide when pre-clearance fumigation is needed, (3) instituting a traceability system so that each shipment can be traced back to the grower, (4) monitoring gall midge progression in the field, and (5) training extension officers and farmers. The hot pepper case illustrates the importance of multi-institutional, farmer-to-consumer strategies for a successful IPM program involving horticultural export crops. More than 400 farmers were assigned a traceability number to participate in the program during 2003.

Other Pre-inspection Programs

Pre-inspection programs are also beginning in Africa. The growth in commercial agriculture in many African countries, including non-traditional peri-urban horticultural crops, has resulted in increased pesticide use. Horticultural crops, produced in Mali after subsistence crops, can be exported to Europe during the winter months to provide an important source of supplementary income to producers. As markets develop abroad, and food safety standards continue to tighten domestically and internationally, environmental quality laboratories (EQLs) are needed to satisfy market requirements for safe foods by testing for residues. In Mali, the IPM CRSP joined with local agencies to develop IPM programs to manage disease and insect pests while reducing pesticide use on vegetables such as green beans that are exported to the EU (see Figure 11-3, page xxiii). IPM strategies were developed, farmers trained through farmer field schools, and technical support and equipment was provided to develop an EQL for residue analysis. These investments will hopefully make it possible to develop a quality assurance program in Mali that meets the stringent requirements of horticultural import markets in Europe. These relatively new efforts in Mali show how African nations, which have historically applied fewer pesticides than other countries, are increasingly forced to address pesticide residue issues.

Conclusions

These examples illustrate: (1) the need to institute pre-inspection programs that include both farm-level IPM and post-harvest quality-control mechanisms if a country hopes to reduce pesticide residues and remain competitive in international markets for horticultural products; (2) the need for public/private partnerships to facilitate adoption and documentation of appropriate pest control procedures; and (3) the benefits of cooperation between public agencies in exporting and importing countries to facilitate pre-inspection. The Guatemalan snowpea and the Jamaican hot pepper cases illustrate the potential for IPM research combined with stringent pre-inspection programs to help facilitate market access. The Guatemalan case also demonstrates the difficulty of instituting widespread pre-inspection programs to meet stringent guidelines when thousands of small farmers are involved. Market requirements may eventually force a shift toward more structured marketing channels for horticultural export crops to meet quality/safety guidelines. If farmers fail to meet them, they will be excluded.

Therefore smaller producers will likely be forced over time to grow in size, to produce under contract, or to join a marketing cooperative.

Acknowledgments

The authors thank Jorge Luis Sandoval, and Luis Calderon, IPM CRSP field technician and coordinator, respectively; Ing Danilo Dardón, ICTA; Karina Illescas, IPM CRSP; Ing. Luis Caniz, APHIS-Guatemala; Ing. Edgar Santizo, Snowpea committee, AGEXPRONT; and Douglas Ovalle, FAS-USDA, Guatemala.

References

FAOSTAT. 2004. <http:\\apps.Fac.org/default.isp>

Institute of Medicine, National Research Council. 1998. Ensuring safe food from production to consumption. Washington, D.C.: National Academy Press.

Julian, J.W. 1999. An assessment of the value and importance of the quality assurance policies and procedures to the Guatemalan snowpea trade. M.S. Thesis, Purdue University, Horticulture and Landscape Architecture.

Julian, J.W., and G.H. Sullivan. 1998. Trade development in non-traditional agricultural export crops: Guatemala's future market competitiveness and sustainability. Purdue AES No. 15877, Proceedings of the Third Central American IPM Seminar, Guatemala City, Guatemala.

Julian, J.W., G.E. Sánchez, and G.H. Sullivan. 2000. An assessment of the value and importance of quality assurance policies and procedures to the Guatemalan snowpea trade. *J. International Food and Agribusiness Marketing* 11(4): 51-71.

Julian, J.W., G.H. Sullivan, and G.E. Sánchez. 2000. Future market development issues impacting Central America's nontraditional agricultural export sector: Guatemala case study. *American Journal of Agricultural Economics* 82(5): 1177-1183.

Lamport, P.P. 1999. Development of IPM techniques for the control of leaf miners on snow peas. M.S. Thesis, Purdue University, Horticulture and Landscape Architecture.

McDowell, H., and S. Martinez. 1994. Environmental, sanitary and phytosanitary issues for western hemisphere agriculture. Western hemisphere/WRS-94-2/June 1994, pp. 73-80.

Murray, D.L., and P. Hoppin. 1992. Recurring contradictions in agrarian development: Pesticide problems in Caribbean Basin non-traditional agriculture. *World Development* 20(4): 597-608.

Norton, G.W., G.E. Sanchez, D. Clarke-Harris, and H. Kone Traore. 2003. Case study: Reducing pesticide residues on horticultural crops, Brief 10 in *Food Safety in Food Security and Trade*, ed. L.J. Unnevehr. International Food Policy Research Institute 2020, Focus 10, Washington, D.C.

Programa de Proteccion Agricola y Ambiental. 1995. Informa final programa de pre-inspection fitosanitaria de melon temporada 1994-95, Julio de 1995 (unpublished report).

Sánchez, G.E., G.H. Sullivan, S.C. Weller, C. MacVean, R. Pérez, R.N. Williams, and L. Calderón. 1998a. IPM CRSP Technical Assistance response in solving the Guatemala snowpea quarantine due to leafminer. Purdue AES No. 15875, Proceedings: Third IPM CRSP Symposium, Blacksburg, Virginia, USA.

Sánchez, G.E., S.C. Weller, G.H. Sullivan, L. Calderón, and ? **Caniz.** 1998b. Establishment of an APHIS-approved pre-inspection program for snow peas in Guatemala under IPM CRSP (unpublished report).

Sandoval, J.L., G.E. Sánchez, G.H. Sullivan, S.C. Weller, and C.R. Edwards. 1999. Effect of strip-cropping on yield and leafminer (*Liriomyza huidobrensis* B.) infestation in snow peas. IPM CRSP Sixth Annual Report 1998-1999, Virginia Tech, Blacksburg, Va.

Sandoval, J.L., G.E. Sánchez, S.C. Weller, et al. 1997. Integrated pest management in snow peas and sugar snaps (*Pisum sativum*) in Guatemala. Proceedings of the Second IPM CRSP Symposium. Guatemala City, Guatemala.

Sullivan, G.H., G.E. Sánchez, S.C. Weller, and C.R. Edwards. 1999. Sustainable development in Central America's non-traditional export crops sector through adoption of integrated pest management practices: Guatemalan case study. *Sustainable Development International*, London, England, launch edition.

Sullivan, G.H., G.E. Sánchez, S.C. Weller, C.R.Edwards, and P.P. Lamport. 2000. Integrated crop management strategies in snowpea: A model for achieving sustainable NTAE production in Central America. *Sustainable Development International*, Autumn 2000 edition: 107-110.

Thrupp, L.A., G. Bergeron, and W.F. Waters. 1995. Bittersweet harvests for global supermarkets: Challenges in Latin America's agriculture export boom. New York: World Resource Institute, p. 202.

U.S. General Accounting Office. 1998. Food safety: Federal efforts to ensure the safety of imported foods are inconsistent and unreliable. GAO/RCED-98-103, p. 64.

U.S. Department of Commerce, Bureau of the Census. 1998. U.S. Imports history: Historical summary 1993-1997 (unpublished data).

Weller, S.C., G.E. Sánchez, C.R. Edwards, and G.H. Sullivan. 2002. IPM CRSP success in NTAE crops leads to sustainable trade in developing countries. *Sustainable Development International*, sixth edition, 135-138.

— IV —
Evaluating Strategic IPM Packages

– 12 –
Evaluating Socio-Economic Impacts of IPM

George W. Norton, Keith Moore, David Quishpe, Victor Barrera, Thomas Debass, Sibusiso Moyo, and Daniel B. Taylor

Introduction

Economic, social, and environmental impacts of IPM are felt by producers, by household members, and by society at large. IPM programs can influence pest control costs, the level and variability of production and income, and the health of pesticide applicators. They can affect food safety, water quality, and the long-run sustainability of agricultural systems. These effects may be felt unevenly by region, farm size, income level, gender, and consumers versus producers.

IPM programs *can* have these effects, but the question remains: what effects *have* specific IPM programs had or what effects *will* they have on one or more of the various factors and groups mentioned above? Answers to these questions are important to producers, to those involved in recommending IPM strategies, and to those who fund IPM research and extension programs. Practical assessment methods are needed to provide credible answers without absorbing an inordinate share of an IPM budget.

A variety of IPM evaluation methods have been applied, and the methods and issues addressed have broadened considerably from the initial field and farm-level budgeting of IPM alternatives. Recent studies have considered risk effects, pest-practice dynamics, and aggregate impacts of IPM programs across regions, gender, and other social dimensions. Health and environmental effects have also been considered, as discussed in Chapter 13. The purpose of this chapter is to briefly describe some basic methods for evaluating economic and social impacts and to suggest what may be a workable evaluation protocol for a participatory IPM program. It describes methods, presents a protocol, and then provides examples based on analyses completed on the IPM and Peanut CRSPs.

Overview of Evaluation Methods[1]

IPM programs are implemented to meet multiple objectives. User objectives include improved profitability, reduced production and income risk, and improved health and safety of workers, among others. Societal objectives often relate to environmental and health concerns associated with pesticide use (see Chapter 13), meeting export standards (see Chapter 11), raising incomes, keeping food prices low, and helping disadvantaged farmers. Because of multiple economic and social objectives and the user/household versus societal level distinction, more than one evaluation technique is usually required. It can be helpful to divide the IPM impact-evaluation problem into three basic components: (1) assessing IPM adoption, (2) identifying economic and social impacts, and (3) assessing health and environmental effects. This chapter focuses on the first two components, with the third one left to the next chapter. The adoption component can be divided into: (a) defining an IPM adoption measure and (b) assessing the degree of adoption. The economic and social/gender evaluation can be split into: (a) assessing user/household-level economic effects and (b) assessing aggregate, market, or societal level effects. The section below summarizes the most common methods used to accomplish these components.

Assessing Adoption

IPM programs may involve several individual practices or strategies, but measuring the impacts of an IPM program can require an aggregate assessment of the program. Therefore, defining what is meant by IPM and determining its degree of adoption are frequently the first steps in an evaluation.

Defining an IPM adoption measure

A commonly understood commodity- and location- specific definition of IPM is needed before the level of adoption can be measured for a specific program. The breadth and complexity of an IPM program often complicates the task of defining what is being measured. PIPM involves local stakeholders in identifying the goals of the IPM program, and can use those stakeholders to define levels of adoption. Levels need defining because producers may selectively adopt IPM practices; therefore IPM adoption can be a matter of degree. In most cases, IPM is defined in terms of the use of

[1] This section draws heavily on the discussion in Norton et al. (2001).

various production practices, although it could be defined in terms of specific outcomes, such as improving some measure of environmental risk. Defining in terms of practices is helpful because the purpose of the evaluation is often to assess the impacts of technologies or practices developed by an IPM program.

Because the desired impacts are usually in terms of improving production, income, or the environment, practices considered to be part of an IPM program should influence those factors. And clearly the practices should be somehow related to the program being evaluated. For example, crop rotation is often cited as an IPM practice, but may or may not be the product of an IPM research program.

Once practices are identified, grouping them to signify levels of IPM adoption (such as low, medium, and high), or relating them to an adoption scale by assigning points to particular practices are common ways to measure the degree of adoption. For example, Rajotte et al. (1987), Napit et al. (1988), Vandeman et al. (1994), and Williams (2000) develop crop-specific definitions of IPM levels based on sets of practices, while Hollingsworth et al. (1992), Petzoldt et al. (1998), and Beddow (2000) developed IPM point systems in the United States. Stakeholders can vary the points they assign depending on their weights on economic versus environmental goals.

There are drawbacks to each approach. Point-system scales can be costly to establish because every technique must be identified and subjectively weighted according to its impact. Whenever a new practice is developed, points for all practices need to be reconsidered. For use in quantitative economic assessments, levels are required, and thus the scale would eventually need to be divided into levels.

If adoption is based on levels representing groups of practices, rather than as a binary indicator or continuous scale, someone must decide how each level is defined. Ideally, levels of adoption would be based on how well the groups of practices met the goals of the IPM program. For example, high adoption would meet nearly all goals of the program, while low adopters would meet few. Such groupings are necessarily somewhat arbitrary.

Representatives of the stakeholder groups can be asked in a participatory process to identify existing strategies or practices available to manage the pest problem(s) within the program boundaries. Once the strategies or practices are identified, they can be grouped into levels or scored on a scale. The more data that can be supplied by experts on the effects of these practices on production or pesticide use, the easier it will be to group or score them. Even with accurate data, the grouping or scoring may vary with the implicit weights attached by stakeholders to income versus other goals.

Assessing the degree of IPM adoption

After creating a measure of adoption, the extent of adoption can be assessed using producer surveys, expert opinion, or secondary data analysis. The method used will depend on the accuracy required, the availability of secondary data, resources available, and whether the study is evaluating IPM practices already adopted or projecting future adoption. Measuring the extent of adoption assumes that one is interested in the aggregate effects of the IPM program. If the interest is only in the relative profitability of IPM practices for making producer-level recommendations, it may be possible to skip the adoption analysis.

The preferred technique for obtaining adoption data on IPM practices already released is to survey potential adopters. Interviews can be conducted in households or in meetings, although, with the latter, subsequent analysis must account for the fact that people self select to attend meetings, which may affect the results. Surveys are relatively expensive and time consuming. In a few cases one may be fortunate to have secondary data that list the adoption of particular IPM practices, but such data are rare. More often, even if there are some secondary data, total or even partial adoption may not have occurred yet. Therefore it becomes necessary to project future adoption. Extension agents or other industry experts may provide reasonable estimates of IPM adoption in the program served, although reliability may vary. Suggestions for predicting future adoption of a technology are found in Alston et al. (1995).

Assessing the determinants of IPM adoption

Even if people adopt IPM practices, it is still necessary to ascertain if a particular IPM program was responsible. The issue of attribution is critical to program evaluation, especially for extension programs. In scientific experiments, a control treatment is designated to determine how much change would take place in the absence of the experimental treatment. Where people are concerned, especially for educational programs, it is often inappropriate or unethical to exclude a control group from the program. Unfortunately, many IPM programs also fail to gather baseline information on the extent to which current practices at the outset of the program already include IPM. In this instance, the only practical way to untangle the effects of natural information diffusion from the added effect of an IPM extension or on-farm research program is to collect survey data and analyze it with regression techniques that allow separation of program effects from other factors that might favor IPM adoption.

A typical regression model to assess the importance of an outreach program would employ a dependent variable representing IPM adoption. This variable could be binary for a single level of IPM, categorical for multiple levels of IPM, or continuous for an index of IPM practices adopted. Explanatory variables would include those factors that would normally be expected to affect technology adoption, such as input and product prices, farmer characteristics, major farm resources, and the physical and institutional setting. Finally, the explanatory variables would include one or more measures of exposure to the IPM program. The significance of the IPM program exposure variable(s) would determine whether observed changes in IPM adoption were attributable to the program or not.

Evaluating Economic Effects

Economic impacts of IPM can be assessed at both the user and societal levels. User effects primarily include changes in costs or profitability, but can include changes in income risk. Societal impacts include changes in the economic welfare of producers and consumers resulting from market-level changes in prices and supply.

User-level Economic Effects

At the user level, IPM adoption implies changes in practices used by producers, with certain practices used more and others less. The result is changes in costs and returns. For example, use of pesticides might decrease while pest monitoring might increase. Decreased use of pesticides would reduce the costs of pesticides, labor, and equipment, while increased monitoring would raise labor costs. Budgets can be used to derive the overall change in net revenue associated with the change in practices. Partial, enterprise, or whole-farm budgets can be constructed. Partial budgets include only benefit and cost items expected to change significantly as a result of the change in practices, while enterprise budgets list all income and expenses (variable and fixed) associated with a particular enterprise. A whole-farm budget includes all enterprises on a farm, and therefore can consider second-order changes in any activity as a result of introducing IPM practices. Partial budgets are the most common and practical type of budget used for assessing IPM impacts. Budgets are constructed for each adoption level (group of practices). A typical partial budget form is presented in Figure 12-1.

By developing a budget for each level of adoption, changes in net revenue can be associated with levels of IPM adoption. Data will be required on inputs, outputs, and their prices. Several options are available for gathering the necessary data. One is to use information on yields and cost of all inputs from on-farm trials conducted by the IPM program. However, the costs and yields for farm-scale IPM adoption may differ from those in the trials. Another option is to conduct a survey of producers in the area targeted by the IPM program using an interview questionnaire. Questions can be included on the same survey used to obtain information on extent of IPM adoption. A sample list of data, to be obtained by farmer recall using such a survey, is provided in Table 12-1.

A third option is to construct enterprise budgets by collecting information on all inputs by operation, preferably by having the farmers collect them in a standard tabular format as they complete each operation, such as land preparation, planting, fertilizing, pest management, cultivating, and harvesting. Data are collected on quantities and prices for inputs such as seeds, fertilizer, pesticides, labor, machinery use, and water and for outputs.

Additions to Net Revenue	Reductions in Net Revenue
Increased Returns:	Decreased Returns:
1. __________ $______	4. __________ $______
2. __________ $______	5. __________ $______
3. __________ $______	6. __________ $______
Total $______ (A)	Total $______ (B)
Decreased Costs:	Increased Costs:
7. __________ $______	10. __________ $______
8. __________ $______	11. __________ $______
9. __________ $______	12. __________ $______
Total $______ (C)	Total $______ (D)
A+C = $______ (E)	B+D = $______ (F)
Change in Net Returns = E - F = $________	

Figure 12-1. Partial Budget Form.

Table 12-1. Example of survey data for economic analysis of IPM.

1. Inputs and outputs	Product yield, and price; percent acreage treated, number of times treated, method of treatment for pesticides, custom spraying, labor for pest management; pest monitoring & prediction information management (pheromone traps, scouting (self or hired), weather models, etc.); beneficial organisms.
2. Extent of IPM adoption	Practices used and percent of acres on which particular practices were used
3. Pest problems and densities	Arthropods, diseases, nematodes, rodents, birds, weeds, elephants.
4. User and farm characteristics	Farm size, acreage of crop, age and education of farmer, gender, years farming, ethnic identification, approximate value of farm products sold, percent of income from farming

Pest population or pressure is measured as well (Figure 12-2). Data are collected on producer and farm characteristics.

A fourth option is to acquire existing enterprise budgets from secondary sources and then to ask scientists or other knowledgeable people to estimate changes in affected categories of the budgets for the various IPM adoption levels. This approach is likely to be the least expensive option.

Testing for significant differences

If any of the first three options are selected, additional statistical analysis is needed to account for statistically significant differences and/or to control systematic differences across farms. Examples of t-tests and F-tests are found in Napit (1986), Norton and Mullen (1994), and Beddow (2000) for calculating differences in mean yields. If field trial data are used, the data will be obtained from experiments in which treatments were randomized, usually in blocks with three or four replications over at least two years. An Analysis of Variance (ANOVA) is conducted with the data to test for significant differences. If the interview-survey or farm-level data collection methods are used, a sample size of at least 30 per sample stratification group is required. For example, if pest management varies by farm

Figure 12-2. Recording field data on plantain experiment.

size group (small, medium, large), then the sample size should be at least 3 X 30 = 90. The cost of interviewing versus farm-level accounting for all costs can differ substantially, and the detailed collection of enterprise data by farm does not necessarily yield more accurate results if outputs and inputs vary significantly from year to year. An interview survey can ask for estimated levels of the most important variables for the past three years to help average out weather-, pest-, or price-induced differences across years. If the interview or farm-level data collection methods are used, regression analysis is the preferred method for testing for significant differences, as it can be used to account for factors affecting yield besides IPM adoption, such as farm size, education, and experience. For example, a yield response equation can be estimated in which dummy variables are included to account for differences in IPM adoption. The t-statistics are calculated for the coefficients on the dummy variables to test for significant differences, while other variables are included in the model to hold constant many of the non-IPM factors affecting yields.

Statistical tests of survey data should account for survey design in order to insure their validity. Many surveys use stratification and/or clustering to enhance the domain of extrapolation of results or to control survey costs. Both approaches have implications for the sample variance of variables collected. Likewise, surveys whose results are intended for future expansion will have probability weights necessary for extrapolation. Probability weights, number of strata, and number of clusters should be factored into survey data analysis to ensure that statistical tests are valid (Deaton, 1997).

Results of budgeting analysis can be used to judge the profitability of practices being developed or recommended to farmers. A second major use of budget information is as an input into a market or societal level assessment of the economic benefits and costs of an IPM program as discussed below. The primary audience in this case may be those responsible for funding the IPM program.

Before turning to market level economic assessments, brief mention should be made of methods for assessing economic risk. Producers who consider adopting IPM strategies may be interested in the degree of risk as well as profitability. Risk may arise from biological, technical, or economic factors. Methods for evaluating risk, such as payoff matrices, and stochastic dominance are summarized in Norton et al. (2001) with examples provided in Reichelderfer, Carlson, and Norton (1984), Greene et al. (1985), Musser et al. (1981), and Moffitt et al. (1983).

Market-level Economic Effects

Market- or societal-level economic impacts are obtained by combining cost, yield, and price changes with adoption estimates, or projections, and with information on responsiveness of supply and demand to price changes. The models used can involve calculation of *economic surplus* changes (as described below). These changes can then be included in a *benefit cost analysis* to account for discounting over time and to facilitate comparisons with other investments.

When widespread adoption of IPM occurs across large areas, changes in crop prices, cropping patterns, producer profits, and societal welfare can occur. These changes arise because costs differ and because supplies may increase, affecting prices for producers and consumers. These changes are illustrated in Figure 12-3. In this figure, S_0 represents the supply curve before adoption of an IPM strategy, and Δ represents the demand curve. The initial price and quantity are P_0 and Q_0. Suppose IPM leads to savings of R in the average and marginal cost of production, reflected in a shift down in the supply curve to S_1. This shift leads to an increase in production and consumption of Q_1 (by $\Delta Q = Q_1 - Q_0$) and the market price falls to P_1 (by $\Delta P = P_0 - P_1$). Consumers are better off because they can consume more

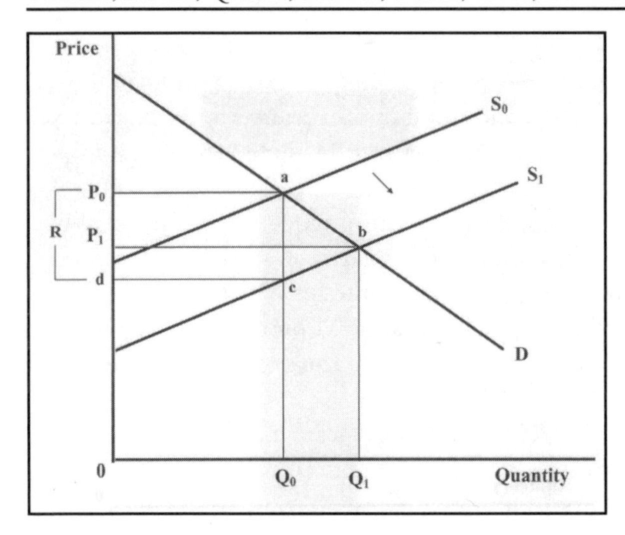

Figure 12-3. IPM benefits measured as changes in economic surplus.

of the commodity at a lower price. Consumers benefit from the lower price by an amount equal to their cost saving on the original quantity (Q_0 x ΔP) plus their net benefits from the increment to consumption. Total consumer benefits are represented by the area P_0abP_1.

Although they may receive a lower price per unit, producers are better off too, because their costs have fallen by R per unit, an amount greater than the fall in price. Producers gain the increase in profits on the original quantity (Q_0 x (R- ΔP)) plus the profits earned on the additional output, for a total producer gain of P_1bcd. Total benefits are obtained as the sum of producer and consumer benefits. The distribution of benefits between producers and consumers depends on the size of the fall in price (DP) relative to the fall in costs (R) and on the nature of the supply shift. For example, if a commodity is traded and production in the area producing the commodity has little effect on price, most of the benefits would accrue to producers. If the supply curve shifts in more of a pivotal fashion as opposed to a parallel fashion, as illustrated in Figure 12-3, the benefits to producers would be reduced. Examples of IPM evaluation using this economic surplus approach are found in Napit et al. (1988) and Debass (2001). Formulas for calculating consumer and producer gains for a variety of market situations are found in Alston, Norton, and Pardey (1995). For example, the formula to measure the total economic benefits to producers and consumers in Figure 12-3, which assumes no trade, is $KP_0Q_0(1 + 0.5Zn)$, where: K = the proportionate cost change, P_0 = initial price, Q_0 = initial quantity, Z = Ke/(e +n), e = the supply elasticity, and n = the demand elasticity. Other formulas would be appropriate for other market situations. It is possible to combine economic surplus models with the results from a Geographic Information System (GIS) to assess the spillover of IPM benefits across regions (see Debass, 2001).

The most difficult aspect of an economic surplus analysis is the calculation or prediction of the proportionate shift in supply following IPM adoption. Cost differences as well as adoption rates must be calculated. The producer surveys, information on cost and yield changes in field trials and other methods discussed above can be used to obtain the information required to estimate the supply shifts. Once changes in economic surplus are calculated or projected over time, benefit/cost analysis can be completed in which net present values, internal rates of return, or benefit cost ratios are calculated. The benefits are the change in total economic surplus calculated for each year, and the costs are the public expenditures on the IPM program. The primary purpose of the benefit/cost analysis is to take into account the fact that benefits and costs need to be discounted, because the sooner they occur the more they are worth. The net present value (NPV) of discounted benefits and costs can be calculated as follows:

$$NPV = \sum_{t=1}^{T} \frac{R_t - C_t}{(1 + i)^t}$$

where: R_t = the return in year t = change in economic surplus
C_t = the cost in year t (the IPM program costs)
i = the discount rate

Changes in economic surplus can also be imbedded in mathematical programming models to further project interregional changes in production following the introduction of a widespread IPM program, or to predict the impacts of IPM following policy changes that encourage or discourage IPM use. In instances where IPM adoption is believed to trigger impacts that touch other sectors in the economy, those impacts may be estimated or projected using sector or computable general equilibrium models.

Aggregate or market level economic effects can be distributed in a variety of ways and have other social and economic effects across and within households. Therefore assessment of welfare effects of an IPM program may not end but rather just begin with evaluation of market-level economic impacts.

Evaluating Social Impacts

Social impacts can be measured in many dimensions: economic (including poverty reduction), health, empowerment, attitudinal, among others. Economic and health dimensions are related to the degree to which IPM strategies are available to and suitable for producers with farms in various size classes, in advantaged or disadvantaged regions, and in different income brackets. Therefore, who adopts IPM and how soon is important. Any technology, including IPM, has impacts across different types of households within the farm and non-farm communities, and impacts within the household, including gendered effects. Measuring these effects can involve both quantitative and qualitative methods.

Quantitative Methods

The simplest method for arriving at basic social impacts is to use the same methods described above for measuring aggregate level economic impacts and to subdivide the benefits and costs by region, income level, farm size, urban consumers versus rural producers/consumers, laborers versus farm owners, etc. Fortunately, the economic surplus approach lends itself to such distributions. In some cases it can be helpful to use geo-referenced data to define homogeneous regions in a GIS. IPM adoption and cost differences can be included to shift out supply curves differentially by region, by farm size, and so forth (Binswanger, 1977; Scobie and Posada, 1978).

Regression analysis can also be used to identify who is likely to adopt IPM, differentiated by farm size, household characteristics, etc. It can be used as well to estimate the effects of IPM on household income, health, or other factors, taking into account farm size and household characteristics. Data for this type of analysis can be obtained from farm/household-level surveys. If gender analysis is intended, the surveys can be administered to both male and female respondents separately. The advantage of applying a quantitative statistical approach is the ability it gives the researcher to hold other factors constant when assessing the effects of IPM.

Increasingly, farm/household surveys are being conducted by international organizations such as the World Bank. Hence it may not be necessary for a program to conduct its own survey to assess social impacts. Data from these surveys can be combined with results from economic surplus analyses, and with poverty measures such as those developed by Foster et al. (1984) to assess the impacts of an IPM program on poverty in a specific country.

An example of such an analysis for a groundnut IPM strategy in Uganda is provided in Moyo (2004).

Qualitative Methods

Qualitative methods are particularly useful for obtaining in-depth assessments of IPM impacts at the household level, because they allow respondents to indicate not only what they are doing but why. They facilitate follow-up questions and group discussions. A common qualitative approach used in developed and developing countries alike is the focus group method. Focus groups are assembled to include members from typical households and structured questioning is used to ensure that all participants feel free to speak and to react to the comments of others. Focus groups can be stratified by gender or other personal or social characteristic. Given the time involved in running a focus group discussion, the approach usually cannot be administered in a way that allows sufficient observations for statistical analysis. But the method provides complementary information to that obtained from such analysis. The method also can be used to help structure the categories for the quantitative analysis.

Another common qualitative technique is the participatory appraisal mentioned in earlier chapters. Much of the one-on-one elicitation of information in a PA is very useful for providing in-depth answers and reasons why an IPM program may affect particular family members and groups in society in specific ways. It too can help structure the more quantitative assessments.

Impact Assessment Examples from the IPM CRSP

IPM impact assessments have been undertaken in Ecuador, the Philippines, Bangladesh, Uganda, and Mali, among other countries in recent years. Most have involved economic evaluations, but some have included social and gender assessments. The following brief examples from the Philippines, Ecuador, Bangladesh, and Uganda briefly illustrate the types of impact analyses that can be completed.

Philippines

Philippine farmers near San Jose, Nueva Ecija, apply insecticides to eggplant twice a week to control fruit and shoot borer. The IPM CRSP tested a series of alternative treatments, including spraying once per week, once every two weeks, once every three weeks, no spraying, and removing damaged fruits and tips at different intervals. An example of a partial budget constructed for one of the treatments in comparison to farmers' practice is shown in Table 12-2. Similar budgets were constructed for each combination in order to arrive at the recommended practice (which turned out not to be the one with the highest yield).

Ecuador

The potato IPM program in Ecuador, undertaken by the national agricultural research institute (INIAP) with the assistance of the International Potato Center and the IPM CRSP, has focused on three primary pests: late blight, Andean potato weevil, and Central American tuber moth. An assessment was made of the economic impacts of a specific potato variety with some resistance to late blight, a set of IPM practices to manage Andean weevil, and a set of IPM practices to manage tuber moth (Quishpe, 2001). The variety was released in the early 1990s and the insect IPM programs are currently being implemented; most of the benefits will occur in the future. Therefore the latter two assessments were basically *ex ante*.

Three producing regions were included in the analysis, and consumption was assumed to occur anywhere in the country and, through trade, in other countries. Data were collected on: costs of the research and extension efforts over time; production, consumption, prices, and trade for the past

Table 12-2. Partial budget for applying Brodan on eggplant every three weeks versus twice a week (farmers' practice)

Particular	Value
Added benefits	
Added revenue	P 168,160.00
Reduced cost	
Insecticide use	11,040.00
labor of applicator	8,333.30
Total Added Benefits	187,533.30
Added costs	
Added cost	
insecticide used	1,932.00
labor of applicator	1,458.30
Reduced revenue	130,240.00
Total Added Costs	133,630.30
Net Added Benefits	P 53,903.00

three years; budgeted costs of production in each region with and without the technologies; expert opinion of extension-type personnel on adoption rates; and price responsiveness (elasticities) of supply and demand.

Economic-surplus and benefit-cost analyses were conducted; calculations were performed using the DREAM program available from the International Food Policy Research Institute (IFPRI). Details of the results are available in the thesis by Quishpe and in Barrera et al. (2002), but, briefly, the research and extension on the late-blight resistant (partially resistant) improved variety (Fripapa 99) produced an internal rate of return of 27% and generated an additional $600 per hectare ($1700 versus $1100) compared to the native variety (Superchola), and roughly $50 million in net benefits.

The IPM program for Andean weevil was estimated to save $87 per hectare in the Central region and $42 per hectare in the South. That pest causes less damage in the North, where the Central American tuber moth is a serious problem. The IPM program against the tuber moth in the North was projected to generate net benefits of $62 per hectare.

Bangladesh

The IPM CRSP program began in 1998 in Bangladesh with a focus on vegetables (eggplant, cabbage, gourds, okra, tomatoes) in rice-based systems. Implemented through the Bangladesh Agricultural Research Council (BARC) and housed in the Horticultural Research Center of the Bangladesh Agricultural Research Institute (BARI), several promising technologies have been developed. Among these technologies are soil amendments to control soil-borne diseases in eggplant, grafting to control bacterial wilt in eggplant, improved weeding strategies for cabbage, and germplasm screened for eggplant fruit and shoot borer. An analysis of returns to the soil amendment (eggplant) and weed-management (cabbage) research was conducted as part of an MS thesis (Debass, 2001).

The evaluation commenced by classifying the country into relatively homogeneous growing areas using a GIS. Variables included for the classification included temperature, rainfall, elevation, population density, land tenancy, production of eggplant and cabbage, and administrative districts (Figure 12-1). Data were collected on production, prices, and elasticities by region for each commodity. Estimates were obtained of the likely per unit cost reductions using partial budgets from the IPM CRSP experiments. Probabilities of research success were estimated by the researchers, as the technologies had not yet been released. Expert opinion was also used to project the adoption profiles. The economic surplus models were solved using DREAM and the results simulated over 30 years and discounted at 5%. Net present values of $14–29 million were obtained for the soil amendment experiment and $15–26 million for the weed research spread over the next 30 years. Most of the benefits were spread over 2 of the 4 regions and one region actually lost in net as a result of the research. Such distributional effects are not uncommon, because some areas are better able to adopt the technologies while all areas feel the downward price pressure that can result from the increased production of commodities that are little traded.

Uganda

Peanut production is constrained by the prevalence of several viruses and diseases, the most common being Groundnut Rosette disease, a viral infection first reported in Tanganyika (now Tanzania) as early as 1907. Groundnut Rosette disease continues to be responsible for major losses in peanut production in Africa. Through the auspices of The International Crop Research Institute for Semi-Arid Tropics (ICRISAT), as well as the USAID-funded Peanut CRSP, peanut varieties with resistance to Rosette virus were recently developed and released in Malawi and Uganda, countries with a high incidence of poverty. Benefits of the research that developed the virus-resistant peanut may have significant economic benefits, and may reduce poverty at the margin in these countries. Benefits may result from higher yields, reduced risk, lower production costs per quantity of peanuts produced, lower food prices, and increased marketed surplus, with possible positive effects on household income.

A study was conducted to estimate the overall economic impacts of the research that developed Rosette Virus-resistance peanut in Uganda, focusing on the regions in the two countries where peanuts are most prevalent, and quantifying the effects of research on the livelihoods of the poor (Moyo, 2004). The evaluation methods contained two major steps. The first involved calculating changes in economic surplus that result from adoption of virus-resistant peanut varieties. The second step involved taking those calculated changes and plugging them into Foster-Greer-Thorbecke additive measures of poverty to compute poverty changes. In short, household production, consumption and expenditure data were used to compute poverty indices that permitted poverty decomposition by income group. Realized research benefits from the economic surplus model were then incorporated into the poverty indexes to estimate how households of differing economic profiles moved relative to the poverty line as their incomes were affected by the improved technology.

To enable estimation of the economic surplus changes, interview and other data were gathered in Uganda along with data on production and output prices by region. Yield and cost data under traditional and virus-resistant varieties as well as realized and projected adoption and research costs were collected from breeders at research institutes, extension officers, farmers, and other industry experts. Data for calculation of poverty indexes were obtained from a national household survey that was conducted by the International Food Policy Research Institute (IFPRI). The data sets are extensive, enabling computation of the poverty indexes and providing a picture of other socio economic characteristics of producers.

Adoption of the virus-resistant peanut varieties in Uganda was estimated to result in a 67 percent increase in yield per hectare and a 50 percent increase in per-hectare input costs. Fifteen percent of the households growing peanuts were estimated to have adopted virus-resistant peanuts in the first year an improved variety was released (2002). Adoption is expected to continue to increase in the future until 50 percent of peanut growers adopt.

The net present value of economic benefits for the period from 2001–2015 was projected at $47 million at a 5 percent discount rate. Changes in poverty rates were then calculated under alternative assumptions about adoption by farmers in various income strata. A Probit model was used with the farm-household data to predict adoption by income strata.

The economic surplus results indicated that adoption of Rosette-resistant peanut seed would result in income derived from peanuts increasing from 75 and 81 percent depending on the future rate of adoption. To estimate the impact of this income change on welfare, three poverty measures were computed for peanut-producing households before and after the adoption of the improved seed. These measures showed a reduction of the poverty rates of approximately 1.5 percent for the region, a significant impact for a single technology.

Conclusions

IPM impact assessment is crucial for making meaningful recommendations to farmers, for demonstrating the value of IPM programs, and for assessing who will adopt so that programs can be tailored to audiences to obtain consistency with program goals. These assessments can be quantitative and qualitative and focus on economic, social, and gender goals. The methods available for impact assessment are relatively cost-effective, although the more data-intensive quantitative ones aimed at calculating factors such as reductions in poverty rates, require enough resources that they may only be conducted periodically in specific locations. Others such as partial budgeting should be routinely applied in virtually every IPM program before recommendations are made to farmers.

References

Alston, J.M., G.W. Norton, and P.G. Pardey. 1995. *Science Under Scarcity: Principles and Practice for Agricultural Research Evaluation and Priority Setting.* Ithaca, New York: Cornell University Press.

Barrera, V.H., D. Quishpe, C. Crissman, G. Norton, and S. Wood. 2002. Evaluacion Economica de la Aplicacion de la Tecnologia de Manejo Integrado de Plagas y Enfermedades (MIPE) en el Cultivo de Papa en la Sierra de Ecuador. INIAP Boletin Tecnico No 91, Quito, Ecuador, Enero.

Beddow, J. 2000. Protocols for assessment of Integrated Pest Management Program. M.S. Thesis, Virginia Tech.

Binswanger, H.P., and J.G. Ryan. 1977. Efficiency and equity issues in *ex ante* allocation of research resources. *Indian Journal of Agricultural Economics Research*, 32, no. 3 (July/September): 217-231.

Deaton, A. 1997. *The Analysis of Household Surveys: A Microeconomic Approach to Development Policy*. Baltimore: Johns Hopkins.

Debass, T. 2001. An economic impact assessment of IPM CRSP activities in Bangladesh and Uganda: A GIS Application. M.S. Thesis, Virginia Tech.

Foster, J., J. Greer, and E. Thorbecke. 1984. A class of decomposable poverty measures. *Econometrica*, 52:3: 761-766.

Greene, C.R., R.A. Kramer, G.W. Norton, E.G. Rajotte, and R.M. McPherson. 1985. An economic analysis of soybean Integrated Pest Management. *American Journal of Agricultural Economics*, 67, no. 3: 567-572.

Hollingsworth, C.S., W.M. Coli, and R.V. Hazzard. 1992. Massachusetts Integrated Pest Management guidelines: Crop specific definitions. *Fruit Notes*, Fall, 12-16.

Kovach, J., et al. 1992. A method to measure the environmental impact of pesticides. *New York's Food and Life Sciences Bulletin*, Number 139. Geneva, New York: Cornell University, New York State Agricultural Experiment Station.

Masud, S.M., and R.D. Lacewell. 1985. Economic implications of alternative cotton IPM strategies in the United States. Texas Agricultural Experiment Station, Information Report No 85-5, College Station, July.

Masud, S.M., R.D. Lacewell, E.P. Boring, and T.W. Fuchs. 1984. Economic implications of a delayed uniform planting date for cotton production in the Texas rolling plains. Bulletin 1489, Texas Agricultural Experiment Station

Moffitt, L.J., L.K. Tanagosh, and J.L. Baritelle. 1983. Incorporating risk in comparisons of alternative pest management methods. Forum: *Environmental Entomology*, 12:4: 1003-1111.

Moyo, S. 2004. The economic impact of peanut research on the poor: The case of resistance strategies to control peanut viruses. M.S. thesis, Virginia Tech.

Mullen, J.D. 1995. Estimating environmental and human health benefits of reducing pesticide use through Integrated Pest Management. M.S. Thesis, Virginia Tech.

Mullen, J.D., G.W. Norton, and D.W. Reaves. 1997. Economic analysis of environmental benefits of Integrated Pest Management. *Journal of Agriculture and Applied Economics*, 29 (2): 243-253.

Musser, W.N., B.V. Tew, and J.E. Epperson. 1981. An economic examination of an Integrated Pest Management production system with a contrast between CV and stochastic dominance analysis. *Southern Journal of Agricultural Economics*, 13:1: 119-124.

Napit, K.B. 1986. Economic impacts of Extension Integrated Pest Management programs in the United States. M.S. Thesis, Virginia Tech.

Napit, K.B., G.W. Norton, R.F. Kazmierczak, Jr., and E.G. Rajotte. 1988. Economic impacts of Extension Integrated Pest Management programs in several states. *Journal of Economic Entomology*, 81:1: 251-256.

Norton, G., and J. Mullen. 1994. Economic evaluation of Integrated Pest Management programs: A literature review. Virginia Cooperative Extension Pub. 448-120, Virginia Tech, March.

Norton, G.W., J. Mullen, and E.G. Rajotte. 1996. *A primer on economic assessment of Integrated Pest Management*, ed. S. Lynch, C. Greene, and C. Kramer-LeBlanc. Proceedings of the Third National IPM Symposium/Workshop, Washington, D.C., USDA-ERS Misc Pub No. 1542.

Norton, G.W., S.M. Swinton, S. Riha, J. Beddow, M. Williams, A. Preylowski, L. Levitan, and M. Caswell. 2001. Impact assessment of Integrated Pest Management programs. Economic Research Service, U.S. Department of Agriculture.

Petzoldt, C. 1998. Elements of processing sweet corn IPM in New York (draft 5/98). Online Source Accessed: May; URL: http://www.nysaes.cornell.edu/ipmnet/ny/vegetables/elements/ei98/prswcorn98.html

Quishpe, D. 2001. Economic evaluation of changes in IPM technologies for small producers to improve potato productivity (Translation from Spanish title). Undergraduate thesis, Central University, Quito Ecuador.

Rajotte, E.G., R.F. Kazmierczak, G.W. Norton, M.T. Lambur, and W.E. Allen. 1987. The national evaluation of Extension's Integrated Pest Management (IPM) programs. Virginia Cooperative Extension Service Pub. 491-010, Blacksburg, Virginia, February.

Reichelderfer, K.H., G.A. Carlson, and G.W. Norton. 1984. Economic guidelines for crop pest control. FAO Plant Production and Protection Paper No 58, FAO, Rome.

Scobie, G.M., and R.T. Posada. 1978. The impact of technical change on income distribution: The case of rice in Colombia. *American Journal of Agricultural Economics* 60, no. 1 (February): 85-92.

Vandeman, A.J., J. Fernandez-Cornejo, S. Jans, and B. Lin. 1994. Adoption of Integrated Pest Management in U.S. agriculture. USDA/ERS, Agric. Info. Bull. No. 707, Washington, D.C., September.

Williams, M.B. 2000. Methodology of an IPM impact assessment: Development and application of a protocol in Michigan tart cherries. M.S. Plan B paper, Department of Agricultural Economics, Michigan State University.

– 13 –

Evaluating the Health and Environmental Impacts of IPM

Leah Cuyno, George W. Norton, Charles Crissman, and Agnes Rola[1]

Introduction

One of the primary motivations for developing and implementing IPM programs is the concern about health and environmental effects resulting from the use or misuse of pesticides. Evidence of the pesticide threat to human health, as well as the tradeoffs between health and economic effects, has been documented in studies in the Philippines (Rola and Pingali, 1993; Pingali, Marquez, and Palis, 1994; Antle and Pingali, 1994; Pingali and Roger, 1995; Cuyno, 1999) and in Ecuador (Crissman, Cole, and Carpio, 1994; Crissman, Antle, and Capalbo, 1998). Many pesticides commonly sold in developing countries are hazardous Category I and II chemicals that are banned or have restricted use in developed countries (Pingali and Roger). Pesticide misuse is common (Tjornhom et al., 1997). In addition, human exposure to pesticides occurs through many routes, not only during application and ingestion, but through indirect means such as entering recently sprayed fields or washing clothes of pesticide applicators.

IPM potentially reduces pesticide use and hence can have health and environmental (HE) benefits, but little empirical work has been completed to estimate those benefits, even in developed countries. Such estimation is difficult because assessing the physical or biological effects of alternative levels of pesticide use under IPM is a challenge, and because most of the benefits are not measured in the market. And, in some countries or regions,

[1] The authors thank Scott Swinton, Susan Riha, Daniel Taylor, Dixie Reaves, Ed Rajotte, Serge Francisco, and Sally Miller for helpful comments and suggestions at various stages of the research. The administrative support of S.K. De Datta and S.R. Obien during the research is also greatly appreciated.

people may not be aware of hazards posed by pesticides (Antle and Capalbo, 1995).

A few studies do suggest possible approaches for measuring HE costs and benefits associated with IPM. Kovach et al. (1992) compared the environmental impacts of traditional pest management strategies with IPM strategies, using a scoring system to consider effects on farmers, consumers, farm workers, and ecology. They derived an environmental impact quotient (EIQ) by pesticide. Higley and Wintersteen (1992) used contingent valuation (CV) to assess the value to farmers of avoiding environmental risks caused by pesticides. They considered effects of pesticides on surface water, groundwater, aquatic organisms, birds, mammals, beneficial insects, and humans (acute and chronic toxicity). Subsequent studies by Owens et al. (1997), Mullen et al. (1997), and Cuyno (1999) used CV analysis to evaluate impacts of pesticides and of IPM. Contingent valuation is controversial, particularly due to the hypothetical nature of the questions used to obtain willingness-to-pay for something like reduced pesticide risk, but steps can be taken to minimize biases. CV has the advantage of being potentially applicable for valuing a broad set of environmental effects.

An IPM program can have numerous health or environmental (HE) effects. Approaches that might be used to evaluate these effects vary greatly in their complexity and the types of information provided. Perhaps the simplest way to document changes that are suggestive of HE effects is to estimate changes in use of pesticide active ingredients (AI). If a more complete assessment is desired, it is necessary to identify pesticide risks to various aspects of health and the environment. These risks might then be summarized in a scoring model that weights and sums risks to various HE categories. Health and environmental benefits can also be scored by using dollars as the unit of measure and combining information on (1) the effects of IPM adoption on pesticide use, (2) reduction in HE risk levels due to IPM, and (3) society's willingness to pay to reduce pesticide risks. Because most HE benefits are not valued in the market, one of a limited number of non-market assessment techniques used by economists could be applied to arrive at an aggregate estimate of net benefits. An alternative approach would be to calculate the external costs directly associated with pesticide use, for example the cost of cleaning up a polluted stream.

In this chapter, we discuss briefly various methods that might be used to evaluate the HE effects of an IPM program, and present a summary of a Philippines study that applied a CV approach.

Methods for Evaluating HE Benefits of IPM[2]

Measuring HE benefits of IPM presents two primary challenges. First, there is a need to assess the physical and biological effects of pesticide use that occur under different levels of IPM. Second, because the economic value associated with HE effects is not priced in the market, it is difficult to know how to weight the various effects. Because of these challenges, the simple estimation of reductions in pesticide a.i. is often used to arrive at a proxy for HE effects, under the assumption that, with pesticides, less is generally better than more. However, because HE effects of pesticides differ dramatically by type of active ingredient (in terms of toxicity, mobility, and persistence), as well as by the method and time it is applied, even a simple accounting of active ingredients (a.i.) should include information on what, how, and when they were applied.

Non-economic Impact Indicators

Reporting on pesticide a.i. indicates environmental effects but does not measure them. To measure them, a model(s) is required to formally link pesticide use to various aspects of health and the environment and then to weight those aspects. Two basic types of pesticide-environmental models are available for impact assessment: location (field)-specific and non-location-specific models. Location-specific models such as GLEAMS or CINDEX (Teague et al., 1995) require detailed field information such as soil type, irrigation system, slope, and weather, and produce information on the fate of chemicals applied. Non-location-specific models usually require information on the pesticides applied and the method of application, and produce indicators of risks by HE category as well as weighted total risk for the pesticide applications. The high data and time requirements (and hence cost) of the location-specific models usually preclude their application except in specialized situations. They can, however, provide useful information on tradeoffs between HE effects and income effects for pest-management practices.

One of the most common non-location-specific indicators of environmental and health impacts of pesticide use is the environmental impact quotient (EIQ) developed by Kovach et al. (1992). The EIQ uses a discrete ranking scale in each of ten categories to arrive at a single rating for each pesticide active ingredient (a.i.). The categories include toxicity to non-

[2] The discussion of methods in this section is a summary of information provided in Norton et al. (2001).

target species (birds: 8 day LC_{50}; fish: 96 hr $LC_{50;}$ and bees), acute dermal toxicity (measured by rabbit or rat LD_{50}), long-term health effects, residue half-life (soil and plant surface), toxicity to beneficials and groundwater/ runoff potential. The EIQ formula groups the ten categories into three broad areas of pesticide action: farm worker risk, consumer exposure potential, and ecological risk. The EIQ is then defined as the average impact of a pesticide a.i. over the three broad areas of action, generally reported as a single number.

Penrose et al. (1994) have proposed a Pesticide Index (PI) as an alternative non-location-specific index. The PI is composed of two individual indices: the Potential for Residues Index (PRI) and the Value Index (VI). PRI calculations are based on the activity of the pesticide (A), the site of application (S), the timing of application (T), and the persistence of the a.i. (P). A, S, T, and P are determined based on an integer scale from 1 to 5. For example, insecticide products receive a weight of 1 if they contain 10 or fewer grams of a.i. per liter.

The value index (VI) is an indicator of the "value or importance of the pesticide in any particular crop production system." It is based on the pesticide's efficacy (Ef), cost (Cs), environmental effects (En), mammalian toxicity (Tx), IPM compatibility (Cp), and the availability of alternatives. Each component of the VI is defined on a subjective integer scale from 1 to 5. PRI and VI are calculated as the sum of the values given to their respective components. The pesticide index (PI) is then the sum of the value index and the Potential Residues Index. The PI recognizes that environmental damages (and benefits) from a given pesticide differ according to the environment in which the pesticide is used (e.g., approximate location and time) and on the method of application

Benbrook et al. (1996) developed an indexing method to assess the aggregate risks of pesticide use to mammals. The index considered acute toxicity via LD_{50} values and developed a composite variable — the mammalian toxicity score (Mam Tox Score) to consider chronic effects. In most cases, the Mam Tox Score for a pesticide is composed of the pesticide's reference dose (RfD), cancer potency factor (Q*), and an indicator of the pesticide's WHO classification (A, B, or C).

Another location-specific index, called the Chemical Environmental Index (CINDEX), was developed by Teague et al. (1995) to describe the effects of pesticides on ground and surface water. CINDEX values are defined for individual pesticide-use strategies. Calculations are based on the 96-hour fish LC_{50}, lifetime Health Advisory Level (HAL) value, the EPA

Carcinogenic Risk Category, and the runoff and percolation potential for each pesticide used in the strategy under consideration. The authors also present a second index, the Chemical Concentration Index (CONC). The CONC is based on the percolation and runoff potential of each pesticide AI in a given strategy, its LC_{50} and lifetime HAL values. The authors posit that the CINDEX and CONC are superior to the EIQ because they are calculated for individual production situations. As such, the index values will differ insofar as runoff and percolation levels are affected by localized factors such as soil type and irrigation systems.

There are many other indexing schemes, including the Environmental Harm Coefficient (Alt, 1976), Environmental Impact Points (Reus, 1998), and PERSIST (Barnard, 1997). Other types of pesticide-indexing methods are described by Levitan et al. (1995), vanderWerf (1996), and Teague et al. (1995).

Using a relatively complex algorithm to assess the environmental effects of pesticides does not remove the subjectivity of the resulting indicators of risk. While both the CINDEX and CONC rating schemes rely on the relatively complex GLEAMS simulation model to derive site-specific percolation and runoff potentials (chemical loadings), the analyst still must specify the relative importance of surface and groundwater (Teague et al. weighted both equally). Thus, the resulting rating is not value free.

Each of the above indexing schemes is a member of a set of assessment and decision-making models known as *scoring models.* Many objections have been raised about how scoring or index-type models have been constructed or implemented, but the major objection that has been raised is with respect to the subjectivity of the weightings and to their tendency to manipulate units of measurement that are not compatible. For example, the meaning of the EIQ is somewhat unclear because of the addition, subtraction, and multiplication of such variables as toxicity, leaching potential, and pesticide half-lives, followed by weighting of categories for aggregation. Often, the figures obtained from scoring models are not readily interpretable and may not be reconcilable with other decision criteria.

In assessing the usefulness of the non-location-specific indexes, it is important to recognize that they perform two tasks. One is to identify the risks of pesticides to the individual categories of health and the environment, such as groundwater, birds, beneficial insects, humans, and so forth. The second is to aggregate across categories through a weighting scheme. Both are challenging tasks. The first is challenging because it is difficult to identify mutually exclusive categories, especially ones where there are data.

The categories in most of these models contain a mixture of non-target organisms (e.g., humans, birds, aquatic organisms, beneficial insects, wildlife) and modes of exposure (e.g., groundwater, surface water). There is overlap, for example between surface water and aquatic organisms. Aggregating across categories is challenging because of the need to weight categories. Most of the differences in the index approaches, then, are due to differences in the categories chosen, the weighting system applied, and the degree of localized detail included in the analysis. One means around the weighting issue is to weight categories with dollars as the unit of measure and willingness to pay for risk reduction as the weights, as discussed below.

Monetary Valuation

When a good is traded in a market, its value is easily determined. In the absence of market distortions, the market price provides an indicator of its value because it captures a great deal of information: consumer willingness to pay, producer cost, scarcity, and so on. Unfortunately, there is usually no market price for health effects and environmental damage. Therefore alternative monetary valuation techniques have been developed, for example contingent valuation and hedonic price analysis, in an attempt to proxy for market prices. Other methods for estimating health and environmental values include a) averting expenditures to avoid exposure to HE risks, b) cost of remediation of HE problems, and c) costs associated with treating illnesses and compensating lost work time due to illnesses induced by HE problems.

While it may seem strange to attempt to place a dollar value on health and the environment, there are several advantages in doing so. Money is a familiar indicator of value and hence easily interpretable. Problems of comparing unlike units are avoided when aggregating across categories or when comparing benefits of a program to its costs. Both monetary and non-monetary indicators attempt to summarize a great deal of information using subjective weights. However, monetary techniques may be less subjective because people assign weights relative to other known values or market prices.

Essentially two approaches to monetary valuation of health and environmental effects have been used. The first focuses on costs and the second on willingness to pay. Health and environmental cost approaches attempt to directly measure the cost of environmental damage or illness. Pimentel and colleagues, for example, have several times used this approach in a crude manner to assess environmental costs associated with pesticides

(Pimentel et al., 1978, 1980, 1991). More careful studies have been completed on the cost of illness and productivity losses due to pesticides in the Philippines and Ecuador by Pingali et al. (1994), Antle and Pingali (1994), and Crissman et al. (1998). Data were collected on pesticide use, demographics, and ailments that were possibly related to pesticide use for pesticide applicators and other relevant individuals. Medical doctors were used for the health assessments. The costs of ailments were regressed on pesticide use, demographics, and other variables so the cost of illness related to pesticides could be estimated. Additional regression models were used to estimate the relationship between labor productivity and health problems related to pesticides.

An alternative cost approach is based on the cost of remediating environmental damage. For example, if pesticides render a supply of groundwater undrinkable, then the cost of remediation is the cost of treating the water to make it potable. When used as an IPM impact-assessment tool, the cost of remediation technique might be used to determine the difference between the total cost of correcting the environmental problem with and without the program. This difference would be a measure of the program's environmental benefit. For any given program it is unlikely that the actual cost with and without the program will be observed, and thus the cost of remediation for at least one of the situations will be hypothetical. Costs of remediation measures tend to be quite high.

Another alternative cost-based approach is the averting expenditure technique. This approach seeks to measure the added costs necessary to avoid exposure to a source of environmental risk. For example, Abdalla et al. (1992) calculated the extra cost of purchasing drinking water in order to avoid drinking contaminated groundwater. Averting expenditures tend to underestimate total environmental impacts, because they focus only on avoiding human exposure.

Willingness-to-pay techniques are widely used in the environmental economics literature to place values on environmental goods or amenities. Hedonic pricing techniques, for example, infer the willingness to pay for environmental amenities from the prices of other goods based on the characteristics of those goods. The approach has been used to value safer pesticides by Beach and Carlson (1993).

A more common approach for valuing IPM has been the contingent valuation (CV) approach. This approach generally employs a survey to collect data on people's willingness to pay (WTP) to receive a benefit or their willingness to accept compensation (WTA) for a loss. Thus the CV can be used to determine values for commodities not traded in the market. A CV survey would typically ask respondents how much they would be willing to pay for a given improvement in an environmental asset, or how much they would be willing to accept in compensation for a given degradation of an environmental asset. In the context of IPM, respondents might be asked to indicate how much they would be willing to pay to reduce the risk of pesticides to various categories of environmental assets. The WTP data could later be linked to pesticide-use data to arrive at a value for a change in pesticide use. Higley and Wintersteen (1992), Mullen et al. (1997), Cuyno (1999), and Swinton et al. (1999) each provide examples of using a CV for such an assessment.

While the CV is a theoretically valid method for determining the value of non-market goods, it is subject to several potential biases that might affect the results (Pearce and Turner, 1990). The more serious biases may arise due to the way the survey is designed or administered, or to the hypothetical nature of the questions. The former can be minimized to some extent, but the latter is particularly hard to eliminate. In an actual market, individuals suffer costs from inaccurate decisions. However, with a CV survey, they face no penalty for an inaccurate valuation; therefore, there may be little motivation to consider all the ramifications of their decisions. If questions are phrased as realistically as possible, hypothetical bias may at least be reduced.

Contingent valuation is one of the few procedures available for cost-effectively estimating environmental costs associated with pesticide use (or environmental benefits of IPM if pesticide use declines). The procedure has been used for roughly 25 years in other settings to estimate non-market costs or benefits, and some progress has been made on survey design and other means of minimizing biases. It is likely to remain a controversial procedure because of the potential for bias, but as a weighting technique it is likely to contain fewer biases than the other indexes mentioned above.

The study by Mullen et al. (1997) illustrates how to use the CV approach to evaluate health and environmental effects of an IPM program. That study first used secondary data to classify each relevant pesticide as high, moderate, or low in risk for the pesticide's effect on groundwater, surface water, acute and chronic human health, aquatic species, avian and mammalian species, and non-target arthropods. Thus, each pesticide was classified by its effect in each of 8 categories and the level of its effect (low, moderate, and high), resulting in 24 risk/environmental classes for pesticides. The effect of the IPM program on pesticide use was estimated using regression analysis. The value of pesticide reduction was estimated using a contingent valuation survey (CVS), the results of which were combined with the estimated reduction in pesticide use attributable to the program to derive the value of the program's health and environmental benefits. Many of the steps would be the same as those needed if alternative index methods were used, as discussed above. The difference is that the HE risks are weighted by the monetary values from the CV survey.

In summary, there are several basic steps to applying a CV approach to valuing HE impacts, including: a) identifying pesticide risks to the environment, b) assessing effects of IPM on pesticide use, c) estimating willingness to pay to reduce risks, and d) calculating risk reductions as a result of the IPM program, and applying willingness to pay estimates to those reductions. Note that if one substituted the word "weights" for the words "willingness to pay to reduce risks", the certain indexing approaches would use essentially the same steps. If one wanted to conduct a quick assessment and consider only changes in pesticide active ingredients, one would simply apply the second step. Each of these steps is illustrated in the Philippines case study below.

Example: Evaluating HE Effects of Vegetable IPM in the Philippines

An economic evaluation of the environmental benefits of the Philippine vegetable IPM program was conducted, with a focus on onion (Cuyno, 1999). The study grouped the steps required for the analysis into three basic categories: a) assessing effects of IPM on HE risks posed by pesticides, b) determining society's willingness to pay to reduce those risks, and c) combining the risks and willingness to pay. These steps are briefly summarized below, and details can be found in Cuyno (1999).

Assessing risks

Assessing risks involved: a) classifying the environment into relevant impact categories, b) identifying risks posed by individual pesticide-active ingredients to each category, c) defining the degree of IPM adoption, and d) assessing the effects of IPM adoption on pesticide use. Environmental categories included the types of non-target organisms affected — humans (chronic and acute health effects), other mammals, birds, aquatic species, and beneficial insects. Previous studies (Higley and Wintersteen, 1992; Mullen et al., 1997) have also included categories for mode of transmission such as surface and groundwater, but these latter categories were excluded for fear of double-counting (i.e., fish live in surface water).

Risks posed by specific pesticides applied to onions in the Central Luzon of the Philippines were assessed by assigning one risk level for each active ingredient for each environmental category using a rating scheme described by Cuyno. Hazard ratings from previous studies were used as well as toxicity databases such as EXTOXNET. Both toxicity and exposure potential were considered in assigning risks for each of 44 pesticides. An overall eco-rating score was then calculated with IPM adoption and without IPM adoption. The difference represents the amount of risk avoided due to the program. The formula for the eco-rating was: $ES_{ij} = (IS_j) \times (\%AI_i) \times (Rate_i)$, where ES_{ij} is the pesticide risk score for active ingredient i and environmental category j, %AI is the percent active ingredient in the formulation, and $Rate_i$ is the application rate per hectare (see Figure 13-1, page xxiii).

The onion IPM program evaluated had been in existence for only five years. Therefore most of the IPM techniques developed in the participatory research program had just been released to farmers, with little adoption yet beyond the local village where the research took place. Therefore, an interview survey of 176 growers in the broader region was conducted to assess farmers' willingness to adopt the IPM practices. Each practice was described to them in the questionnaire, and they were asked, if a particular IPM practice were to be available to them next year, would they adopt it. During the subsequent environmental impact analysis, sensitivity analysis was conducted with the results of the adoption survey, as the hypothetical nature of the questions posed casts some doubt on the accuracy of the responses. In addition, Logit models were used to project adoption in the broader region by taking the responses for the 176 farmers and regressing their willingness to adopt specific technologies on a set of socioeconomic characteristics. The actual characteristics for the broader population were then used in the model to predict the probability of adoption in the region as a whole. Details are provided in Cuyno.

Expected reduction in pesticide use as a result of adopting the IPM technologies was based on experiments conducted in farmers' fields through research supported by the IPM CRSP. This program had developed IPM practices to control a small red insect (*Thrips tabaci*), weeds (especially *Cyperus rotundus*), cut worms (*Spodoptera litura*), soil-borne diseases (particularly *Phoma terrestris* or pink root), and nematodes (*Meloidogyne graminicola*). By the time the environmental assessment was conducted, components of the IPM program were released (or near release) for *Thrips*, weeds, cutworms, and *Phoma*. These components included practices that reduced the usage per hectare of specific insecticides for *Thrips* and for cutworms by 50 percent for those farmers who adopt them, herbicides by 65 percent, and fungicides for pink root by 25 percent.

Willingness to pay

To place a monetary value on the environmental benefits of the onion IPM program, estimates were needed of society's willingness to pay (WTP) to avoid pesticide risks to the five environmental categories. WTP values were obtained through CV using a survey of 176 randomly selected farmers in Nueva Ecija district. Strategies were employed to minimize strategic, information, starting-point, vehicle, and hypothetical biases. Following Van Ravensway and Hoehn (1991) and Owens et al. (1997), an approach was used to minimize hypothetical bias by simulating a market (buy and sell exercise) for a good that is similar to another good familiar to the respondents. Farmers were asked to provide willingness-to-pay values for different formulations of their favorite pesticides. Five formulations were offered, one that avoids risk to each of the five environmental categories. For example, farmers were asked whether they would purchase their most commonly used pesticide, reformulated to avoid risk to human health, at a series of prices (in 50-peso increments) higher than its existing price. The estimates of willingness to pay to avoid pesticide hazards to the various environmental categories were then adjusted downward by 30 percent to reflect the fact that the pesticides in the local area were applied 70 percent on onions during the dry season, and 30 percent on other crops, principally rice and other vegetables.

Combining pesticide hazard and willingness-to-pay information

Reductions in pesticide hazards due to implementation of the five IPM practices were calculated by multiplying the risk score for each pesticide by the percent active ingredient, and then multiplying this result by the application rates per hectare, with and without the IPM practices. The percent reduction in this eco-rating hazard was multiplied by the willingness-to-pay value for each category to arrive at an economic benefit per person. Aggregate benefits were obtained by multiplying the per person value by the number of people in the local region. However, the resulting aggregate benefit is an underestimate for two reasons. First, the IPM technologies are likely to spread beyond the local region, even if adopted at a lower rate. Second, benefits to consumers of the onions beyond the local area are assumed to be zero.

Results

The *Thrips* control practices developed in the IPM program involved reduced frequency of applying pesticides with the active ingredients Clorpyifos + BPMC. The weed control IPM practices reduced the use of Glyphosate, Fluazifop P-Butyl, and Oxyfluorfen. The cutworm IPM practices reduced the use of Lambdacylhalothrin, Cypermethrin, and Deltamethrin. The disease control IPM practices reduced the use of Benomyl and Mancozeb. The risk scores for these pesticides are presented in Table 13-1.

Risk scores were calculated for an additional 34 pesticides (not presented due to space limitations) because calculation of the percent reduction in environmental hazards required consideration of all active ingredients, not just the ones used for onions and the particular pests addressed.

The scores assigned to each pesticide active ingredient (by category) were combined with usage data to arrive at an overall ecological rating for each pesticide, as noted above. Eco-ratings with the IPM program took into account the adoption projections, which ranged from 36 percent for an integrated weed/insect/disease control practice to 94 percent for an IPM practice that reduced herbicide treatment from two sprays to one spray. The eco-ratings were reduced from 60 to 64 percent as a result of the IPM program, depending on the environmental category (Table 13-2). These reductions represent the percent pesticide risk avoided.

Table 13-1. Risk scores for onion pesticides applied in the study area/affected by IPM practices (5 = high environmental risk; 0 = no toxicity)

	Environmental Category				
Active Ingredient	Human	Animals	Birds	Aquatic	Beneficials
Benomyl	4	4	3	5	5
Mancozeb	3	3	3	5	5
Fluazifop-P-Butyl	4	4	0	5	5
Glyphosate	4	4	3	3	3
Oxyflourfen	4	4	1	5	5
Chlorpyrifos + BMPC	3	3	5	5	5
Cypermethrin	3	3	5	5	5
Deltamethrin	4	4	3	4	5
Lambdacyhalothrin	3	3	3	4	5

Table 13-2. Eco-ratings with and without the vegetable IPM program

Category	Type of pesticide	Eco-ratings without IPM	Eco-ratings with IPM	Aggregate % risk avoided
Human health	Herbicide	323	114	
	Insecticide	405	142	64
	Fungicide	20	15	
Beneficial insects	Herbicide	332	117	
	Insecticide	456	180	61
	Fungicide	28	21	
Birds	Herbicide	122	43	
	Insecticide	405	161	60
	Fungicide	23	17	
Animals	Herbicide	323	114	
	Insecticide	405	142	64
	Fungicide	20	15	
Aquatic species	Herbicide	331	117	
	Insecticide	358	132	62
	Fungicide	27	20	

The farmers' willingness to pay to reduce pesticide risk to various environmental categories is presented in Table 13-3. These values ranged from 551 to 680 pesos per cropping season (40 pesos = $1 U.S.) and were within a reasonable range given household budgets in the area, although perhaps a little on the high side. The values were adjusted downward to reflect the use of the pesticides on other crops and were multiplied by the

Table 13-3. Willingness to pay for environmental risk avoidance and economic benefits

Category	Mean WTP (pesos per season)	WTP adjusted for % of pesticides on onions	Economic benefits (WTP adjusted by % risk avoided)
Human health	680 (219)[1]	476	305
Beneficial insects	580 (197)	406	248
Birds	577 (200)	385	231
Animals	621 (198)	434	278
Aquatic	551 (210)	404	250

[1] Standard deviation is in parentheses.

percent risk avoided to arrive at the benefits per person per season (1312 pesos). These benefits represent more than 6 million pesos for the roughly 4600 local inhabitants in the five villages (rural neighborhoods) where the IPM program is centered, or about $150,000 (Figure 13-2). While not all household or community members (i.e., children or non-farmers) may value the environmental and health benefits as much as the farmers who were interviewed, the $150,000 is likely to be an underestimate of the total benefits to the community because the IPM practices are likely to spread well beyond the 10-kilometer radius of the five local villages. In addition, farmers received direct economic gains from the lower production costs associated with these practices. For example, the savings in direct pesticide costs for some of the IPM practices are roughly twice the environmental benefits, based on separate calculations.

Conclusions

IPM programs developed to help solve pest problems while minimizing pesticide use are potentially a win-win situation. They may raise agricultural productivity while reducing environmental damage and improving health. Public support for IPM programs may be justified on their produc-

Figure 13-2. Farmers' willingness to pay to reduce pesticide risks in the Philippines totaled $150,000 for the five villages surveyed.

tivity effects alone, but in many cases, a significant share of the benefits may be missed if HE gains are ignored.

An assessment of HE benefits may also assist in designing pesticide regulations. Any institutional arrangement that serves to reduce pesticide use by the indicated amount would generate these benefits. The analysis in the Philippines illustrates that it is possible to estimate HE benefits in a relatively low-cost manner in a developing country using a farmer survey. The use of CV for such an analysis may in fact have an advantage in a developing country like the Philippines where many of the beneficiaries are farmers who are familiar with pesticide use. It is often said that growers in such a country may be less aware of the dangers of pesticides. However, to the extent that they have direct experience with the chemicals and yet represent a significant proportion of the rural population, they are a logical group to survey. In developed countries, farmers are a much smaller percent of the population; hence, a CV survey on pesticide risk would need to focus mostly on consumers who may have very little experience with farm chemicals.

For more detailed assessment of human-health costs of pesticide use, the type of approach used by Crissman et al. in Ecuador may be preferred to CV analysis. Measurement of clinical symptoms provides a more precise assessment of acute health problems and allows one to determine who within the household is being affected. In summary, there are various approaches for evaluating HE benefits of an IPM program. While not as straightforward to apply as are methods for assessing productivity benefits, and somewhat more costly to implement, these approaches can be applied in situations where concerns over HE effects are paramount.

References

Abdalla, C.W., B.A. Roach, and D.J. Epp. 1992. Valuing environmental quality changes using Averting Expenditures: An application to groundwater contamination." *Land Economics* 68 (2): 163-169.

Alt, K.F. 1976. An economic analysis of field crop production, insecticide use and soil erosion in a sub-basin of the Iowa River. M.S. Thesis, Iowa State University.

Antle, J.M., and S.M. Capalbo. 1995. Measurement and Evaluation of the impacts of agricultural chemical use: A framework for analysis. In *Impact of Pesticides on Farmer Health and the Rice Environment,* ed. P.L. Pingali and P.A. Roger. Norwell, Mass.: Kluwer Academic Publishers.

Antle, J.M., and P.L. Pingali. 1994. Pesticides, productivity, and farmer health: A Filipino case study. *American Journal of Agricultural Economics* 76, 418-430.

Benbrook, C.M., et al. 1996. *Pest Management at the Crossroads*. Yonkers, N.Y.: Consumers Union.

Barnard, C., et al. 1997. Alternative measures of pesticide use. *Science* 203: 229-244.

Beach, D., and G.A. Carlson. 1993. Hedonic analysis of herbicides: Do user safety and water quality matter? *American Journal of Agricultural Economics* 75: 612-623.

Crissman, C.C., J.M. Antle, and S.M. Capalbo. 1998. Economic, environmental, and health tradeoffs in agriculture: Pesticides and the sustainability of Andean potato production. Norwell, Mass.: Kluwer Academic Publishers.

Crissman, C.C., D.C. Cole, and F. Carpio. 1994. Pesticide use and farm worker health in Ecuadorian potato production. *American Journal of Agricultural Economics* 76: 593-597.

Cuyno, L.M. 1999. An economic evaluation of the health and environmental benefits of the IPM program (IPM CRSP) in the Philippines. Ph.D. Thesis, Virginia Tech.

Higley, L.G., and W.K. Wintersteen. 1992. A novel approach to environmental risk assessment of pesticides as a basis for incorporating environmental costs into economic injury level. *American Entomologist* 38: 34-39.

Kovach, J., C. Petzoldt, J. Degni, and J. Tette. 1992. A method to measure the environmental impact of pesticides. *New York's Food and Life Sciences Bulletin* No. 139, New York State Agriculture Experiment Station, Cornell University, Ithaca, N.Y.

Levitan, L., I. Merwin, and J. Kovach. 1995. Assessing the relative environmental impacts of agricultural pesticides: The quest for a holistic method. *Agriculture, Ecosystems and Environment* 55: 153-168.

Mullen, J.D., G.W. Norton, and D.W. Reaves. 1997. Economic analysis of environmental benefits of Integrated Pest Management. *Journal of Agricultural and Applied Economics* 29(2), 243-253.

Norton, G.W., S.M. Swinton, S. Riha, J. Beddow, M. Williams, A. Preylowski, L. Levitan, and M. Caswell. 2001. Impact assessment of Integrated Pest Management programs. Economic Research Service, U.S. Department of Agriculture.

Owens, N.N., S.M. Swinton, and E.O. Van Ravenswaay. 1997. Farmer demand for safer corn herbicides: Survey methods and descriptive results. Michigan Agricultural Experiment Station, Michigan State University. Research Report 547, East Lansing, Mich.

Pearce, D.W., and R. K. Turner. 1990. *Economics of Natural Resources and the Environment*. Baltimore: Johns Hopkins University Press.

Penrose, L.J., W.G. Thwaite, and C.C. Bower. 1994. Rating index as a basis for decision making on pesticide use reduction for accredidation of fruit produced under Integrated Pest Management. *Crop Protection* 13: 146-152.

Pimentel, D., et al. 1980. Environmental and social costs of pesticides: A preliminary assessment. *Oikos* 34 (2): 126-140.

Pimentel, D., et al. 1978. Benefits and costs of pesticide use in U.S. food production. *Bioscience* 28 (12): 772-790.

Pimentel, D., et al. 1991. Environmental and economic impacts of reducing pesticide use. *Bioscience* 41 (6): 403-409.

Pingali, P.L., and P.A. Roger. 1995. Impact of pesticides on farmer health and the rice environment. Norwell, Mass.: Kluwer Academic Publishers.

Pingali, P.L., C.B. Marquez, and F.G. Palis. 1994. Pesticides and Philippine rice farmer health: A medical and economic analysis. *American Journal of Agricultural Economics* 76, 587-592.

Reus, J.A. W.A. 1998. An environmental yardstick for pesticides: State of affairs, ed. E. Day. Proceeding of the Integrated Pest Management Systems Workshop, Chicago.

Rola, A.C., and P.L. Pingali. 1993. Pesticides, rice productivity, and farmers' health: An economic assessment. International Rice Research Institute, the Philippines and World Resources Institute, Washington, D.C.

Swinton, S.M., N.N. Owens, and E.O. van Ravenswaay. 1999. Health risk information to reduce water pollution. Pp. 263-271 in *Flexible Incentives for the Adoption of Environmental Technologies in Agriculture*, ed. E. Day. Boston: Kluwer Academic Publishers.

Teague, M.L., H.P. Mapp, and D.J. Bernardo. 1995. Risk indices for economic and water quality tradeoffs: An application to Great Plains agriculture. *Journal of Production Agriculture* 8(3): 405-415.

Tjornhom, J.D., K.L. Heong, V. Gapud, N.S. Talekar, and G.W. Norton. 1997. Determinants of pesticide misuse in Philippine onion production. *Philippine Entomologist* 1(2): 139-149.

VanderWerf, H.M.G. 1996. Assessing the impact of pesticides on the environment. *Agriculture, Ecosystems and Environment* 60 (1996).

van Ravenswaay, E.O. and J.P. Hoehn. 1991. Contingent valuation and food safety: The case of pesticide residues. Department of Agricultural Economics Staff Paper No. 91-13, Michigan State University, East Lansing.

– 14 –
Gender and IPM

Sarah Hamilton, Keith Moore, Colette Harris, Mark Erbaugh,
Irene Tanzo, Carolyn Sachs, and Linda Asturias de Barrios

Introduction

The term *gender* will be used in this chapter to refer to non-biological differences between women and men. Gender affects the adoption and impacts of IPM in myriad ways, as women prove to be both more involved than expected in household decisions concerning pesticide use and less able than are men to obtain access to IPM information and technologies in most of the countries in which the IPM CRSP has worked. Even in regions where women are unlikely to view themselves as "farmers," IPM CRSP researchers have found that women often influence pest-management strategies through their roles in household budget management and decision-making regarding expenditure for pesticides and labor. Generally lacking information about IPM, these women are as likely as their male counterparts to prioritize spending for pesticides. However, in all countries, women shoulder most of the responsibility for family health and are more concerned than are their spouses about the health costs of pesticide use; in diverse settings, women are responsive to information concerning alternative forms of pest management they view to be more healthful. While in most countries women have less access than do men to technical support in IPM, in those areas (primarily African) where women have equal access, they are at least as likely as men to adopt IPM. Thus we view obtaining accurate information about women's roles in the work and decision processes that ultimately result in household pest management choices to be a critical component of participatory IPM research. We also emphasize the elimination of gender barriers in participatory research and in the dissemination of research results through agricultural extension and other forms of outreach.

Gender Differences and Constraints to IPM Adoption

The gender differences that most consistently affect the adoption and impacts of IPM, positively or negatively, include gendered:

- Ownership and control of resources
- Family provisioning responsibilities
- Division of household labor
- Specialized knowledge of crops, pests, and traditional pest-management practice
- Attitudes toward health consequences of pesticide use
- Social roles.

The constraints to IPM adoption that are faced by women around the world are both material and ideological. Even in areas where women provide significant agricultural labor on family farms, they may not be viewed — even by themselves — as "farmers." Time allocation and labor recall studies in Latin America and Asia have found that it is not unusual for women who average as many as eight hours a day in household agriculture to report their occupation as "housewife," the socially more appropriate or self-preferred role definition. Even in areas where local women and men typically view women who work in household agriculture as farmers, women's roles in household agriculture can remain invisible to agricultural researchers and extension workers. Most invisible of all of women's responsibilities in household agriculture are their roles in decision-making. It is not unusual for well-meaning members of research teams and agricultural extension programs to expect that even women who work full-time in household agriculture are merely unpaid family labor and that all technical and other relevant decisions are made by men (Hamilton, 2000).

The lack of social recognition as farmers has resulted in low levels of formal membership in organizations that offer technical and marketing assistance for the horticultural export crops targeted by the IPM CRSP in countries such as Guatemala, Jamaica, and the Philippines, though informal participation in marketing is higher. Social exclusion of women from male-oriented groups in Africa and Asia has been ameliorated by the presence of female extension workers in some areas. IPM CRSP programs have been intentionally made available to equal numbers of male and female farmers in Uganda, where women are widely viewed to be farmers. In Mali, where women are also viewed as farmers, equal numbers of female and male farmers have recently been included among IPM CRSP collaborators in Farmer Field Schools.

Access to labor is also expected to be a constraint to adoption of IPM for women farmers. IPM is often more labor-intensive, as well as information-intensive, than the use of chemical pesticides. In many settings, women have less control than do men over household labor and also less money with which to hire labor. Moreover, women may not control their own labor in the most patriarchal settings. However, as will be seen below, women appear to be able to overcome this constraint, at least in the IPM CRSP countries, where — in all but one — women own much less land than men do. Women make land-use decisions, including what crops to plant, on their own land and on land held jointly with their husbands. In the Latin-American case studies below, women even make land-use decisions, together with their husbands, on land held independently by the men. These decisions can have important implications for adoption of export horticultural crops and the subsequent use of pesticides and other pest-management choices in production of these crops. However, women tend to control less land independently than do men in systems in which land-use decisions are likely to be taken independently by women and men, rather than jointly. This disparity can mean that women are differentially constrained from access to credit and other factors of production in export markets.

Women find ways to address many of the production constraints with which they are faced, and the adoption of IPM can help women to improve their positions in both households and markets. As will be seen below, women not only have obligations to provision their families with food and income from agriculture, but they also use the value of their labor and knowledge to improve their positions within intra-household economic reciprocities. Women can raise the value of their labor through developing new crops, new marketing outlets, and skills. They also raise the value of their labor through developing alternative agricultural wage- and labor-exchange opportunities, as they have done through women's production associations and work groups in many settings (see chapters in Rocheleau, Thomas-Slayter, and Wangari, 1996). In "two-tier" farming systems such as those in Africa where women farm both independently and on household fields, women are particularly likely to seek new opportunities in horticultural crops, as they are more likely to be able to recruit labor for these labor-intensive crops than to have access to the greater extensions of land required for traditional grain crops. In areas where women are not perceived to be farmers, women are likely to seek "non-traditional" outlets for their productive labor, including new crops in whose markets they hope to play a formative role. In all of the IPM CRSP research countries, women are involved in the small-scale production of commercial crops, most for export. Since IPM has been demonstrated to help producers supply a high-quality horticultural export product, access to technical support in IPM becomes a matter of gender equity as well as economic efficiency.

An additional gender difference that can affect IPM adoption and impact is the specialized knowledge that results from gendered divisions of labor in household agricultural production and marketing. Women and men often specialize in different crops or in different tasks within one cropping system. Traditional knowledge of pests that is associated with crops or with performing particular tasks is often strongly gendered. In some settings (e.g., Bangladesh), women's traditional knowledge of integrated pest management is in danger of being lost. Many studies have documented the fact that women often have knowledge of a greater number of species than do men and that women are somewhat more likely to practice "integrated" agro-ecology as they piece together livelihoods from often marginal spaces in farms, forests, and coastal areas (see Sachs, 1996; and volumes edited by Sachs, 1997, and by Rocheleau, Thomas-Slayter, and Wangari, 1996). Women may perceive pest problems that men do not, and vice versa. For this reason, as well as to increase adoption rates and for gender equity, it is important to include women in IPM participatory research (see Figure 14-1, page xxiv).

Women's Roles in IPM in Ecuador, Guatemala, Jamaica, the Philippines, Bangladesh, Mali, and Uganda

While the differences between men and women in terms of labor, access to resources, and specialized knowledge can be striking, it is perhaps even more striking that despite these distinguishing features, in IPM CRSP case studies, there are no gender differences in commitment to pesticide use. While levels of pesticide use differ among crops, pesticide regimes per crop do not differ by gender. This similarity holds for women and men involved in green beans in Mali, mixed crops in Uganda, onions in the Philippines, commercialized vegetables in Bangladesh, potatoes in Ecuador, snow peas and broccoli in Guatemala, and peppers in Jamaica. In Mali, a mixed-gender control group of "non-collaborating" green bean farmers, 97.1 percent of women used insecticides vs. 92.1 percent of men. In Uganda, when all crops were taken together, there were no overall gender differences in the proportion who used pesticides nor in the number of applications. This is a critical point because women are as committed as men to pesticide use, but have much less access to technical support in IPM.

Women's participation in decisions concerning the choice of pesticide, quantity of pesticide, and application schedule is much greater than most researchers anticipated in some regions of Latin America, the Caribbean, and Asia, where women's roles in commercial agriculture are less visible than in Africa. In random-sample household surveys, IPM CRSP researchers found that women choose and/or purchase pesticides in 70 percent of all farm households in three indigenous communities in the Central Ecuadorean highlands; 50–75 percent (depending on community) of all households in three communities in Jamaica (Hamilton, Schlosser, Schlosser, and Grossman, 1999); 40–60 percent of farm households (depending on whether survey respondent was male or female) in two regions in Central Luzon, Philippines; and 33 percent of all farm households in three indigenous communities in the Central Guatemalan highlands. Ecuador also provides a case study in the importance of ethnicity in determining whether women will have decision-making roles in agriculture. Women had much less decision authority in two ethnically mestizo areas, one of which was similar in terms of all structural features and cropping systems, when compared with the indigenous communities in the Central Highlands. In Africa, where women are expected to be independent farmers, it was not so surprising that up to one hundred percent of sampled women farmers make these decisions, depending on who actually applies the pesticides (women who successfully recruited their husbands to apply the pesticides often left the choice of chemical to the men). Across regions, it was not unusual for the person who applied the pesticides to have a role in selection of the chemical. But application was not the only route to decision-making regarding pesticides.

Women's roles in household budget management provided entrée into decisions regarding pesticide use and, ultimately, pest-management strategies in case studies from Ecuador, Guatemala, and the Philippines. In nearly 90 percent of households in indigenous communities studied in Ecuador and in the Philippines, women controlled pooled household budgets derived from all household incomes. Women also prioritized spending for pesticides but were more interested than men in saving money by reducing pesticide dependence. IPM is considered risky by these women, but they are eager or willing to learn more about alternative approaches. Women were involved in decisions concerning budget allocation for pesticides even in households in which they do not participate in agricultural labor or consider themselves to be farmers. In Guatemala, where household budgets are not pooled, women control their own incomes and administer the household subsistence fund into which their husbands and wage-earning children contribute (Hamilton, Asturias de Barrios, and Tevalán, 2001). Nearly half of all male commercial farmers stated that they had to borrow from their wives in order to buy pesticides; the men also revealed that the women were not obligated, and could not be compelled, to provide funds.

Women's attitudes toward pesticides do not differ from men's except in one important regard: women are more concerned about the health costs of pesticide use and abuse. Women are more directly responsible for family health and nutrition, are likely to be concerned for health risks to their own and other women's fetuses, and are concerned about the safety of foods they and their families consume. We have found no evidence that women are "naturally" more eco-friendly and concerned about the physical environmental costs of pesticide use than are men. When calculating trade-offs among economic and environmental utility, women and men are equally likely to feel they must sacrifice environmental utility for immediate livelihood needs. But women are much more likely to be receptive to messages concerning the health and safety of their families, including their husbands, than men are likely to be concerned for their own health and safety. Case studies from Ecuador and Uganda show that when women receive some information about the health risks of pesticide use, they are likely to seek more information and are more likely than men to seek alternative crops or methods of pest management to reduce these risks. Women's slightly higher IPM scores in Uganda were related to their health concerns.

When women have equal access to technical support in IPM, they are as likely to adopt as are men. Genuinely equal access means that information is transmitted in a form in which women can understand it and in an environment in which they can participate freely. In Mali, where women have recently been added into the ranks of "collaborating" IPM CRSP farmers who attend Farmer Field Schools, women's scores on IPM accelerated dramatically and nearly equaled those of male collaborators, some of whom had been attending such sessions over a longer period and most of whom were more familiar with the educational platforms. In Uganda, women and men scored nearly equally on a scale of IPM knowledge and practice.

Informally, women are perhaps somewhat more likely to learn about IPM from social networks than are men. This was the case in Mali, where 21% more women who had never attended a Farmer Field School had discussed IPM practices with Farmer Field School participants than had non-collaborating men, and 10% more of these women had heard of Farmer Field Schools than had the men (although fewer of the non-collaborating women had adopted some of the practices).

Case Studies

The dynamics of gendered IPM adoption and outcomes are highlighted in case studies below. Cases address women's work and the labor-related constraints and benefits of IPM adoption perceived by women; women's control of productive resources and intra-household decision processes associated with IPM; women's attitudes towards pesticide use and IPM; and institutional constraints to women's adoption of IPM and promotion of IPM within household production systems.

Ecuador

The three sites included in the Ecuador study demonstrate variation across regions and ethnicities in women's roles in pest management. In Cotopaxi in the Central Highlands, indigenous women farmers from cash-cropping households participated equally with their husbands in choosing pesticides, in deciding how much to apply, and in the application of toxic chemicals. Women also bore primary responsibility for managing the household budget, including production expenditures, and for maintaining family health. Thus women farmers are the household members most likely to make decisions that affect the health costs of overuse of hazardous chemicals. In the Northern-Highland province of Carchi, where study households were mestizo rather than indigenous, women were much less likely to participate in either the work or the economic decisions that impact pest management, but, like the women of Cotopaxi, were responsible for family health and concerned about the economic and health costs of pesticide use. In the coastal province of Manabí, mestizo women from plantain-producing households appeared to be less involved in agricultural labor and decision-making than were women in the indigenous area of the highlands, but more involved than their ethnic counterparts from the highlands. Baseline data collection demonstrated that, at all three sites, women historically had not benefited from technology transfer concerning agricultural pest management. These cases highlight the importance of determining women's existing and self-desired roles in pest management as early as possible during program planning.

Ecuador, Site 1: Central Highlands, indigenous population

A study was carried out in 1993 and 1998-99 in three indigenous communities in Cantón (county) Salcedo, Cotopaxi Province. Quantitative data were collected from a stratified random sample of male and female household heads in 62 households from a population of around 470 households. Qualitative information was collected through participant observation and informal interviews with 10 families, representing the full range of farm size and agricultural modernization, over a period of 10 months.

Although the population is indigenous, most people speak Spanish rather than, or in addition to, traditional Quichua. The local economy is based on small-scale commercial potato production, supplemented by dairying. On average, households derived 90 percent of cash income from agriculture; half supplemented this income with the non-agricultural earnings of male heads. Around one-fifth of men worked outside the community during 1992–1993, averaging 14 weeks per year. All but two households owned some land; holdings averaged 2.9 hectares (SD 2.3). The highly unequal distribution of land among households was positively skewed; two-thirds owned fewer than three hectares and 83 percent owned fewer than four hectares. However, the gendered division of land ownership was egalitarian (Hamilton, Asturias de Barrios, and Tevalán, 2001). Seventy-seven percent of women and 70 percent of men owned land. The quantity and quality of land owned by women (av. 1.8 hectares, SD 2.5) and men (av. 2.3 hectares, SD 4.6) did not differ significantly (probability of t = .430). Household lands were located at altitudes ranging from 3000 to 3500 meters.

Among the most serious threats to family livelihoods, women and men ranked pest-management problems alongside drastic annual rainfall fluctuations, soil erosion on steep slopes, and crop-killing frosts at higher elevations. Most pesticides were applied to the potato crop, which is highly vulnerable to the fungus late blight (*Phytophtera infestans*) and to the Andean weevil (*Prenotrypes vorax*) (Chrissman and Espinosa, 1993:16). In a three-month period during the dry season of 1993, the average household sprayed each field five times; some households applied once a week (Table 14-1). Application levels were reportedly higher during the rainy season. In 1999, a similar pesticide regime was in place (see Figure 14-2, page xxiv).

Table 14-1. Pesticide use on potato in three indigenous communities, Cantón Salcedo, Ecuador (3-month period, dry season, 1993; N=62)

Most Frequently Applied Pesticides:

Trade Name	Active Ingredient	Class of Chemical	Amt./Ha. Mean
Dithane M-45	Mancozeb	Thiocarbamate	2.5 kg.
Monitor	Methamidophos	Organophosphate	460 cc.
Furadan	Carbofuran	Carbamate	540 cc.
Cosan	Sulfur	Metal/mineral sulfur	2.3 kg.

Frequency of Application:

	Mean	SD	Range
Number of applications per field	5	4	0-17

Women's Participation in Household Pesticide Use:

Apply pesticides .... 71% Select pesticides* 70%

*Self-reported equal or greater participation than husband in decisions concerning selection and quantity of pesticides

Nearly all women were full-time farmers who participated in all agricultural tasks except plowing with draft animals and driving tractors. Women controlled their own labor and often were responsible for organizing household labor and for hiring and supervising additional labor. All household lands were worked jointly by both household heads, and the income from all household production was pooled. There were no "women's crops" or "men's crops." Men and women worked approximately the same number of hours per day. Both men and women reported that the agricultural work of men and women was equal, and men and women agricultural workers earned an equal wage.

Both women and men also reported that agricultural decision-making is a consensual process in which women have an equal voice with their husbands. Eighty-four percent of women reported at least equal participation with their husbands in decisions regarding land use (when and what to plant). Although financial decision-making was a joint process, women bore greater responsibility for all decisions regarding investment of household funds, including production expenditures, as they were regarded as the guardians of household income. Both women and men said that women had the authority to execute this responsibility in 90 percent of households. Women also were generally acknowledged to be decision-making caretakers of family health, and the family members most likely to be responsive to information concerning health and nutrition.

Women applied pesticides and participated equally with their husbands in decisions concerning product and dosage in 70 percent of households (Hamilton, 1998: 158-160; 178-180). As keepers of the household purse, women tended to be more stringent in their assessment of how much the household could afford to spend for pesticides and more demanding in their requirements for product quality. Women were primary decision-makers concerning how much their households could afford to buy. While men usually made the purchases, they did so only with their wives' approval of the outlay and the negotiated sale. Women made these decisions on farms of all sizes, regardless of whether they owned land independently or whether their husbands were absent from the farm (Hamilton, 1998: 254-255). Women were more likely to execute their decision authority in selecting and purchasing pesticides if they also applied pesticides or if their households used larger quantities of the chemicals.

Women's attitudes toward pesticide use reflected both their providership and family health responsibilities. They prioritized spending for pesticides, but were willing to invest their labor and that of other family members in alternatives that would require smaller cash outlays or reduce perceived threats to respiratory and dermatological health. Women also expressed unmet needs for technology transfer in the areas of potato and dairy production. At the time of the study, the only locally-available information on pest management was provided by chemical companies. An NGO provided dairy veterinary extension to a predominantly male group recruited from a male-oriented political group; this same NGO offered extension in family nutrition to a woman's group organized originally by the Roman Catholic church. This gendered division among extension opportunities mirrors patterns observed in the following case from the northern highlands and elsewhere in Ecuador and Latin America.

Ecuador, Site 2: Northern Highlands, mestizo population

The study was carried out in 2000 in the Tulcán, Montúfar, and Espejo cantons of Carchi Province, in Ecuador's far northern highlands bordering Colombia (Barrera et al., 2001). Sixty farm households participated in a quantifiable survey, and additional interviews were conducted to clarify information. In terms of ecology and production system, the population was similar to that of Cotopaxi. The local agricultural economy was based primarily on small-scale commercial potato and dairy production, characterized by high levels of toxic chemical use. In the survey region, 76 percent of farmers owned fewer than three hectares (Barrera and Norton, 1999). Despite these structural similarities, there were striking differences in women's roles in agriculture and in pest management in the two highland settings.

Agricultural labor and decision-making were much more differentiated by gender than in Cotopaxi. In Carchi, men tended to be exclusively dedicated to potato and other crop production. Women participated in sowing, harvesting, and seed selection, but had a more important role in animal and dairy production. Women were not viewed primarily as farmers. Few women applied chemicals or made land-use and pest-management decisions, even if their husbands were away from the farm while working in regional cities. In 86 percent of households, the "family leader" (generally male) or the leader's husband (in cases where the male is absent) bought pesticides. The predominantly male leader or woman's husband hired daily workers in 82 percent of households. In terms of decision-making, predominantly male leaders and women's husbands made land-use decisions in 78-83% of households (depending on the decision domain). The selection of pesticides is made by predominantly male leaders or women's husbands in 87 percent of households. In no case did a married woman perform any of these tasks or make any of these decisions. Joint decision processes involving both husband and wife were not measured. Comments in focus groups suggested that joint decision-making in which women would have equal participation would characterize a minority of households at best.

At the time of the 2000 baseline study, 100 percent of men, but few women, had received training in agriculture (primarily in potato production). Women expressed a desire for technology transfer in IPM and in veterinary science. Women also expressed interest in developing the small-scale manufacture of IPM technologies and in developing alternative crops. Women in Carchi were, like their counterparts in Cotopaxi, concerned about health issues arising from overuse or misuse of pesticides. Men were less concerned about health issues than were their wives, but also desired training in IPM and the health effects of pesticide use.

Ecuador, Site 3: Coastal Region, mestizo population

A study was carried out in 2002 in the El Carmen district of Manabí Province. Male and female household heads were interviewed in some 300 plantain-producing households (Harris et al., 2002). Plantations were small; nearly all were less than 20 hectares and half were five hectares or less. In 32 percent of households, farm management decisions were made by male and female heads together or by women alone (7.4 percent). While this level of decision making by women was less than that observed in the indigenous area in Cotopaxi, it was much greater than in the mestizo highland population of Carchi. Survey responses did not provide a great deal of information about women's labor in plantain production. However, women's knowledge of some plantain cultivation practices was similar to men's. This finding suggests that women may be more involved in production than had been expected.

Ecuador: Explaining regional and ethnic differences

The striking differences in women's control over land and agricultural technology in the two similar highland areas can be explained, at least partly, by ethnicity. In Cotopaxi, women's equal ownership of land is based in traditional indigenous inheritance practice. Women's voice in decision-making derives from both ownership and from traditional preference for bifurcated household headship. There is a deeply-held belief that households with two equally authoritative heads will prosper, both because "two heads are better than one" in terms of expertise, and because such balance provides a moral basis for success (Hamilton, 1998).

In Carchi, on the other hand, the patriarchal gender ideology that has been widely documented in highland mestizo communities appears to hold sway. In focus groups, women maintained that gender inequality in economic decision-making derived from beliefs that men are more responsible for economic providership and thus deserving of inheritance in land than are women. A corollary belief is that men should be respected as the unitary heads of their families who are charged with directing the household's economic activities.

These profound differences in cultural values persist despite generations of participation in commercial potato markets. While our data do not explain variation between the highland and coastal mestizo communities, it has been widely observed that mestizo socio-cultural institutions tend to be more conservative in the highlands than on the coast.) These cases point to the importance of allowing for variation in women's roles even within socio-economically similar groups in a single country.

Guatemala: Nontraditional Export Sector, Central Highlands, Indigenous Population

The roles of women in work on and management of farms and pest-control regimes varies across indigenous populations in Latin America. A case study from Guatemala finds that the women work and manage economic resources in a less gender-egalitarian context than that of indigenous women in highland Ecuador, but that their roles in decision-making related to pest management are greater than resource ownership would predict.

The study was carried out in two predominately Kaqchikel Maya communities located in the municipalities (counties) of Tecpán and Santa Apolonia. Nontraditional export crops — primarily snow peas and broccoli — were first adopted in the early 1980s at one site and in in the late 1980s in the other. Quantitative analysis is based on a 1998 random-sample survey of 141 households, from a population of some 380 households (Asturias et al., 1999), and a follow-up 2001 survey of 214 men and women from a randomized subsample of 113 households (Hamilton, Asturias de Barrios, and Sánchez, 2002). Ninety-four percent of the sampled population self-identified as Kaqchikel Mayan, nearly all bilingual in Kaqchikel and Spanish. Qualitative results are based on interviews and observations carried out in these and neighboring communities in 1998–2001.

The local economy was primarily based in agriculture. Over 80 percent of male household heads reported household agriculture as their primary occupation, and half of all households included members who earned wages as agricultural laborers. Seven percent ran agricultural wholesale businesses. Nonagricultural income sources included services, textile and earthenware artisanry, and storekeeping. Nontraditional agricultural export (NTAE) production characterized 50–64 percent of all households (depending on year). Ninety-seven percent of NTAE producers also planted subsistence crops, compared with 91 percent of households who produced commercial crops only for the domestic market. The distribution of land was highly and positively skewed. On average, households owned slightly less than one-half hectare. One-fifth of households did not own land; an additional three-fifths owned less than 1 hectare; and only 3 percent owned 5 or more hectares.

The work and management of household agriculture were locally perceived to be a primarily male domain, reflecting a traditional division of labor in which men were primarily responsible for agriculture while women earned incomes from marketing agricultural and nonagricultural products, animal production, craft production, and storekeeping. Rather than the pooled household budget under women's administration that was observed in Cotopaxi, here tradition favored separate budgets for household subsistence, managed by women, and for agricultural production, managed by men. However, the results of surveys and qualitative data collection revealed that women were heavily involved in the work of household NTAE production and had much more influence on agricultural decision-making than we had expected (Table 14-2).

Among commercial producers, women worked with crops in 94 percent of households, primarily in planting and harvesting; one-fourth of women were involved in cultivation and one-tenth in land preparation. Women marketed crops in many households and were considered the primary producers of income derived from non-bulk marketing of household crops (i.e., crops sold in regional markets rather than one-time sale to exporters and other bulk buyers) in 16 percent of all households. Women's primary control of non-bulk marketing provided independent income and yielded higher prices than household bulk sales (Hamilton, Asturias de Barrios, and Tevalán, 2001). Women's other sources of income included animal production and storekeeping and other petty commerce; a few women sold agricultural produce in bulk or worked as agricultural market intermediaries.

Women's work in agricultural marketing proved to be an important determinant of their ability to influence household decisions concerning the use of land on the extremely small farms on which livelihoods depended. Women's land-use decision participation is particularly relevant to IPM adoption in this population because the decision to devote land to NTAE

Table 14-2. Women's roles in household nontraditional agricultural export (NTAES), Central Guatemalan Highlands* 1998-2001

Percentage of Women Involved in NTAE Labor and Decision-making	
Household NTAE field labor	94%
Household land-use decisions*	78%
Transfers from household budget to purchase pesticides	46%
Pesticide selection	33%
Application of agrochemicals other than pesticides	20%
Membership in groups receiving IPM technology transfer	3%
Pesticide application	0%

* N = 87, 1998 Household Survey; N = 214, 2001 Household Survey; land-use decision-making was measured in both surveys, with identical results; all other results are from the 1998 survey.

crops results in much higher levels of pesticide use than for other crops and because IPM technology transfer takes place largely through export-oriented cooperatives and companies. Compared with the 57 percent of men who had inherited land, only 22 percent of women held inherited land. Another 29 percent of women had bought land together with their husbands. Yet women made land-use decisions jointly with their husbands or independently in over three-fourths of all households. Multivariate regression analysis indicated that women were more likely to control decisions about what crops to grow on family land if they owned land or if they marketed household crops (Hamilton, Sullivan, Asturias de Barrios, and Sánchez, 2000).

Women did not apply pesticides. Local people voiced beliefs that women were unable to shoulder the heavy backpack sprayers and would be more adept at work requiring manual dexterity, such as harvesting snow peas, and work requiring careful attention to fine detail, such as sorting produce for market quality grades. Yet women participated in pesticide-selection decisions in nearly one-third of all households. And, perhaps as important, women's control of the household subsistence fund provided leverage to influence decisions regarding the purchase of pesticides. In two-thirds of households producing export crops, women and men reported that women shared control of export proceeds. Women administered the household subsistence fund to which husbands and adult children contributed a portion of all earnings. Even though part of this fund was comprised of agricultural proceeds contributed by husbands, the fund was under the wife's control. If a man wanted to tap this fund for money to purchase pesticides, he could not do so without his wife's approval. In the 46 percent of sampled households in which the husband reported he had to borrow from his wife to purchase pesticides, women had considerable leverage to influence the level of pesticide use.

The overuse and misuse of pesticides in this population have been persistent problems, even though IPM CRSP activities in participatory research and technology transfer have begun to reach a large number of producers with encouraging results. In the snowpea sector, pesticide-related product detentions were a recurring problem during the 1990s. Although extension workers involved in IPM CRSP activities stated that women came to many of their programs, the women generally stood in the back and did not participate actively. Very few women were formal members of the production associations that collaborate in IPM activities. Even though women were generally not viewed by extension workers, by husbands, or even by themselves as producers of export crops, they clearly had economic roles that impact IPM adoption.

Uganda: Eastern Region, Iganga and Kumi Districts

In Eastern Uganda, pesticide use levels did not differ between male and female farmers; household heads, whether male or female, were perceived to be dominant decision-makers with respect to the use of pesticides across an array of crops, and married individuals with spouses present reported shared decision-making. Women had equal access to IPM CRSP activities and other extension services and at least equal knowledge of IPM.

The study was carried out in 1995 and 2000 in 8 villages, from which random samples of 25 farmers in each village were selected (Erbaugh, Donnermeyer, Amujal, and Kyamanywa, 2003). Among survey respondents, 52 percent were female. Unlike the studies from Latin America, not all survey respondents were heads of household, nor were women and men conjugal pairs from the same household. Among women, 40 percent were heads of their households. The range of crops included maize, beans, groundnuts, sorghum, and cowpea. Coffee and tomatoes were also grown in some areas, but were not included in measures of crop-management knowledge.

There were no gendered differences in the amount of land planted in crops or in the crops grown by women and men (with the exception of coffee (perceived as a male crop) and tomatoes (grown by women in one district). Both men and women said that land clearing was predominantly male or performed by both genders; planting, weeding, and harvesting were performed by both. Men said that both men and women performed land clearing, while women said men predominated in this domain.

With respect to agricultural decision-making, there were significant differences between women's and men's views of who makes decisions, with women more likely to report that they make decisions alone or with their husbands, while men reported their own control of decisions or that decisions were made jointly with spouses. Overall, a relatively equal decision role for women and men emerged. Twenty-four percent of all respondents reported that men dominate; 26 percent reported that women dominate; and 50 percent reported that women and men make decisions together. With respect to who controls income from agricultural production, there was a less profound difference between male and female accounts. Women were reported to have slightly more control over incomes than over agricultural decisions: 25 percent of all respondents said that men controlled farm incomes; 29 percent said that women controlled these incomes; and 46 percent said that both controlled the incomes. Individuals from households with both spouses present were more likely to report shared labor and decision making.

Women were as likely as men to use pesticides (Table 14-3); the number of applications did not differ by gender — an average of six applications for men and five for women per season. Further, women and men were equally likely to want to increase their level of pesticide use. While women were just as likely to have their fields sprayed, men were more likely to do the applications and to purchase and make decisions concerning the chemicals (Table 14-4).

Early in the IPM CRSP participatory research process, women farmers were targeted for inclusion in program activities. Women and men were equally likely to have access to extension workers' visits. Women had equal knowledge of beneficial insects and pest-control alternatives and were more aware of negatives associated with pesticide use, primarily health costs. Thus, even though women remained as eager as men to increase the level of

Table 14-3. Pesticide use in eastern Uganda, 2000

All (N=200)	MAle (N=96)	Female (N=104)	Total	X^2	phi
- Not Using	37	37	74	.188	.031
- Using	59	67	126		
By District	Iganga N=100)	Kumi (N=100)			
- Not using	56	18	74	30.97**	.394**
- Using	44	82	126		

Degrees of freedom = 1; * p < .05; ** p < .01. Source: Erbaugh et al., 2003.

Table 14-4. Person in the household who makes pesticide-use decisions (pesticide users only), Eastern Uganda, 2000

Pesticide Users Only (N=126)	Male (N=59)	Female (N=67)	Total	X^2	Cramer's V	Phi[1]
- men	44	11	55	47.51**	.614**	
- women	3	32	35			
- both	12	24	36			

X^2 and Cramer's V, Degrees of Freedom = 2; Phi, Degrees of Freedom =1. Source: Erbaugh et al., 2003.

their pesticide use, women's scores were higher than men's on an index of IPM knowledge. There were some gendered differences in crop-management knowledge scores per crop; women had higher scores for groundnuts and cowpea, while men had higher scores for beans. There were no differences for maize and sorghum.

These findings point to the importance of learning about women's roles early in program planning, as did the cases from Latin America. In Eastern Uganda, the IPM CRSP survey in 2000 measured extension access and IPM knowledge five years after activities began. The inclusion of equal numbers of women in program activities significantly raised the number of pesticide-using farmers with exposure to IPM information. Given women's heightened awareness of negatives associated with pesticide use, it is hoped that women will begin to adopt IPM in greater numbers.

Mali: Eastern Zone of the Office de la Haute Vallée du Niger (OHVN), Green Bean Export Sector

In this West-African setting, several striking similarities with the Ugandan case emerge: agricultural and pesticide-use decisions are tied to task rather than to ownership of crop or field; men apply pesticides on their wives' plots; women are as likely to use pesticides as men are and to desire to raise the level of their pesticide use; and women with access to IPM information quickly began to respond positively (in the Malian case, women began to adopt IPM practices). These similarities are all the more striking because of both cultural and agricultural differences between the two regions: women's decision roles and market participation might be expected to be lower or less visible in the heavily Islamic West-Africa setting and in the export market. Further, in Mali, the fact that men did the pesticide application did not result in their greater decision input on their wives' plots. Women appear to have as much control over pest management as their East African counterparts.

The baseline study was carried out in 2000-2001 in three villages: Sanambélé, Koren, and Dialakoroba (Sissoko et al., 2001). The sample was formed of 34 male collaborators in the IPM CRSP farmer field trials and Farmer Field Schools, 38 male non-collaborators, and 34 female non-collaborators. Sharp gender inequities characterize the distribution of human and material capital. Literacy rates were more than twice as high for men (49% for non-collaborators and 41% for collaborators) as for women (21%). Few women owned land, and those who did owned much less than men did. Average garden size was 160 sq.m. for collaborating men, 105 sq.m. for non-collaborating men, and 77 sq.m. for women (mean differences are significant at the .01 level). Most women had to access land through their husbands. Despite these differences, all women and men ranked green beans their most important horticultural crop for both income and consumption. In general, women reported that they decide how to use their income from green beans.

Overall, decision-making appeared to be a function of labor contribution. Labor contributions to the spouse's production, whether by men or women, correlated positively with decision input on the spouse's green-bean plot. Men's and women's labor contributions to their own and their spouses' green-bean plots were measured for 14 tasks. On 13 of the 14 tasks, women's labor increased decision input on their husbands' plots; on 12 of the 14 tasks, men's labor increased their decision input on their wives' plots. Because a larger proportion of men contributed to tasks across the full range of activities, overall men gained more decision input through labor transfers than did women. But this finding also meant that husbands provided their wives with more labor, a primary production constraint.

In terms of pesticide use and decision input, however, the relationship between labor and decision-making did not hold. The fact that a man's labor contribution did not accord him a decision role concerning pesticide application on his wife's plot does not necessarily mean that women retained unadulterated authority over pesticide use, however. Ultimately, the green-bean exporter appeared to be a farmer's most influential partner in pesticide-related decisions: 31.4 percent of farmers said the company provided advice on pesticides; 50.5 percent said that farmers and the company shared in pesticide-application decisions; and 8.6 percent said the company made the decisions. Over two-thirds of all green-bean producers applied insecticides with a sprayer owned by the exporter's agent. Another one-fifth of men owned their own sprayers. While 90 percent of all collaborating and non-collaborating farmers believed that pesticides always increase yields and most farmers have not decreased pesticide use, 15 percent more of collaborating farmers have reduced pesticide use levels than have non-collaborators (the difference is not significant at the .05 level).

In 2000–2001, having access to IPM CRSP activities made a significant difference in pest-management practices, but gender did not. Since no women were involved in IPM CRSP activities in 2000, women could only be compared to other non-collaborators. Among non-collaborating women and men, there was little adoption and no significant gender differences on adoption of neem leaf extract, colored flags, cabbage residues, lonchocarpus, solarization, or burning residues. In the study villages, knowledge of chemicals came from exporters and OHVN, while knowledge of IPM was transferred largely through research channels such as the IPM CRSP activities.

In 2001–2002, the IPM CRSP increased efforts to include women farmers in these activities. Women's Farmer Field Schools were established, and some 179 women gained training in IPM (Sissoko et al., 2002). A 2002 impact analysis of seven villages demonstrated dramatic increases in women's adoption of IPM (Table 14-5) and a narrowing of the gender gap between women and men. In this sample there were larger differences between women and men non-collaborators than in the three-village sample; non-collaborating men tended to adopt more IPM practices than did women (Table 14-6).

Gender workshops conducted by IPM CRSP Mali Site researchers indicated that women were constrained in their participation in the Farmer Field Schools by lack of education, as most were not literate and could not read leaflets and brochures or comply with some recording activities (Sissoko et al., 2002). Subsequent workshops on training methodologies were provided to help alleviate this constraint. Women's productive and reproductive labor requirements and lack of transportation also constrained participation. Given these important constraints, it is not surprising that a gender gap persisted between male and female Farmer Field School participants in IPM adoption. The more remarkable fact is that so many women had adopted part of the package, even though they had only recently begun to participate in training activities and faced more constraints than men

Table 14-5. Practices adopted by IPM CRSP collaborators, Mali, 2002

	Women		Men		Totals	
IPM Practice	no.	%	no.	%	no.	%
Burning plant residues	20	71.4	19	79.2	39	75.0
Neem leaves	19	67.9	20	83.3	39	75.
Well-decomposed manure	21	75.0	22	91.7	43	82.7
Loncocarpus	1	3.6	2	8.3	3	5.8
Colored vaseline traps	5	17.9	10	41.7	15	28.8
Tobacco powder	4	14.3	6	25.0	10	
Cabbage residues	5	17.9	8		13	
Solarisation	3	7.1	7	25.0	8	15.4
Seedbed preparation	15	53.5	12	50.	27	51.9
Seeding method	7	25.	11	45.8	18	34.6

N=52, 28 = women, 24 = men. Source: IPM CRSP Mali Site Researchers

Table 14-6. IPM practices adopted by IPM CRSP non-collaborators, Mali 2002

IPM Practice	Women		Men		Totals	
	no.	%	no.	%	no.	%
Burning plant residues	6	20	12	41.4	18	30.5
Neem leaves	12	40	9	31	21	35.6
Fertilizer	9	30	16	55.2	25	42.4
Loncocarpus	0	0	0	0	0	0%
Colored vaseline traps	1	3.3	4	13.8	5	8.47
Tobacco powder	0	0	0	0	0	0%
Cabbage residues	0	0	2	6.9	2	3.39
Solarisation	0	0	2	6.9	2	3.39
Confection des planches	3	10	14	48.3	17	28.8
Methode de semis	1	3.3	0	0	1	1.69

N=59, women= 30, men = 29. Source: IPM CRSP Mali Site Researchers

did. Women also indicated that pest-management practices that did not involve spraying insecticides were attractive to them because they could implement them independently rather than depending on the labor contributions of their husbands. An important corollary of the workshops was that some men came to a better understanding of their wives' labor constraints and began to contribute more of their own labor to women's production.

The constraints women faced in gaining access to IPM CRSP technology-transfer activities have been mirrored in many settings. By providing separate training for men and women, the CRSP helped women gain access in a cultural setting that required physical separation of women and men. Tailoring the training to women's educational levels would also increase women's adoption of IPM.

Philippines: Central Luzon, Nueva Ecija

IPM CRSP research in Nueva Ecija in 1997 illustrates several methodological issues relating to the collection of information on women's roles in pest management. First, while the project attempted to collect information from all adults in each household who contributed to the family economy, female household heads were under-sampled because they were not identified as "farmers" or individuals who "work on the farm" or as individuals who make "farm-related" decisions. From the female household heads who passed the first of these gateways, a picture emerged suggesting that, like their counterparts in Ecuador, women in Nueva Ecija are budget managers who make highly relevant decisions impacting their households' use of pesticides. This finding formed the basis for methodological revisions and hypotheses for later research.

Both IPM CRSP initial PA results in San José in 1994 and those of a 1997 household survey conducted in San José and Bongabon — rice and onion farming communities in Nueva Ecija, Luzon — suggested that women were involved in decision domains that directly affect pest-management practices (Tanzo, Heong, and Hamilton, 1998). Survey results were considered preliminary because gateway questions involving whether survey respondents were "farmers" or were engaged in field work on the farm resulted in women's being interviewed in only 32 of the 58 households surveyed. Rather than asking to speak with both male and female household heads, researchers asked a household spokesperson to identify all family members who worked on the farm or earned income off farm. Because women who did not work in the fields were not identified among those who "work on the farm," they were not interviewed unless they also earned off-farm income. Men, however, were expected to be decision-making farm managers even if they did not do field work in their own agricultural enterprises.

Of these respondents, nearly 90 percent of men and of women indicated that women are their families' primary financial managers (Table 14-7). Since these reports agreed with other studies from the region showing that household budgets are pooled under the management of the female head (Mehta, Hoque, and Adalla, 1993), we were particularly interested in women's views of the costs and benefits of allocating household funds to the purchase of pesticides. Further, approximately 60 percent of women claimed to be their families' primary pesticide buyers, a figure that proved consistent with earlier PA findings, while the same proportion of men claimed to be the primary buyer on the household survey. Differences between men's and women's reports on pesticide purchase were statistically significant (Table 14-7). The PA also indicated that women were likely to decide when to spray and to select pesticides. Taken together, these preliminary quantitative and qualitative results indicated that women had budget-management roles that would impact their households' pest-management practices, and that this form of participation could characterize households in which women did not perform field labor.

Because women had to clear a second gateway by self-identifying as individuals who made "farm-related decisions," only 11 of 32 female

Table 14-7. Activities, knowledge, and perceptions relating to household pest management practice by gender of household head, Nueva Ecija, Philippines, 1997

	Male Heads N = 54		Female Heads N = 32		
	Mean	SD	Mean	SD	Sig. t 2-tail
Reporting Roles of Both Male and Female Head					
Report Primary Pesticide Buyer is Male Head	63%	49	41%	50	.045
Report Primary Pesticide Buyer is Female Head	26%	44	59%	50	.003
Report Primary Financial Manager is Female Head	89%	32	88%	34	.848
Reporting Role of Self Only					
Will Borrow for Pesticide Purchase	91%	29	94%	25	.627
Spend for Food First	50%	51	47%	51	.782
Spend for Pesticide First	17%	38	25%	44	.354
Divide Equally Between Food and Pesticide	22%	22	22%	42	.971
Reduced Pesticide Use > No Perceived Benefit	39%	49	47%	51	.474
Reduced Pesticide Use > Crop Loss, Damage	33%	48	16%	37	.058
Reduced Pesticide Use > Saving Money	9%	29	19%	40	.245
Knowledge of IPM	44%	50	34%	48	.364
Willingness to Adopt for Onion	100% (N = 25)	0	81% (N = 11)	40	.167

Source: Tanzo, Heong, and Hamilton, 1998.

household heads in the survey responded to questions concerning attitudes and practices related to pesticide use. Although this sub-sample is much too small to be considered representative of the sample or its population, the patterns revealed by the survey provided another set of testable hypotheses and suggested that women in Nueva Ecija resemble their counterparts in Latin America and Africa in prioritizing spending for pesticides, desire to use pesticides, and desire to save money by using less pesticide (Table 14-7). Women were less likely than men to perceive potential crop loss or damage from failure to use pesticides, and they were more likely than men to consider the money-saving benefits of reduced pesticide use. A particularly interesting pattern was that women and men did not differ on their priorities concerning whether to spend first on food or on pesticides. Around half of all respondents would spend for food first, while 25 percent of women and 17 percent of men would spend first on pesticides. Over 90 percent of both women and men stated their willingness to borrow to finance pesticide purchases.

Farmers in one of two areas where the 1997 IPM CRSP survey was conducted belonged to a large export-oriented onion production cooperative. Although the organization was headed by a woman and financial resources were controlled and channeled primarily through women officers and employees, membership was primarily male. Wives of male members were encouraged to attend meetings, either together with husbands or in their stead, and they did so. But few of the farmers collaborating in participatory field trials were women. Thus, in this case it was not surprising that women had less knowledge of IPM than men did and that their stated willingness to adopt IPM for onion production was lower than men's was (Table 14-7).

Conclusions

In the regions studied by IPM CRSP researchers, overall high levels of participation by women in both export and domestic agricultural markets and in pesticide regimes have been documented. Both this finding and the wide range of variation observed in women's IPM-related work, knowledge, and decision pathways lead us to conclude that an understanding of women's roles in pest management, and in household agriculture and economies more generally, is essential in achieving the desired environmental and economic benefits of IPM. Women's participation is key to program success in all phases, from initial reconnaissance and stakeholder identification through iterative planning and implementation, to dissemination of results.

There is no substitute for collecting high quality information that can be disaggregated by gender. These forms of data include the Participatory Appraisals that were analyzed for gender differences and for women's roles and agendas early in the history of each program site. The PAs, with their rich qualitative content, were complemented by large quantifiable household surveys and other surveys. For these surveys to be useful in gender analysis, both male and female heads must be interviewed in all households. Female household heads must be included because, among the women in a household, they may be the most likely to have budgetary control and to participate in land-use decisions, even if they do not work in the fields. Other female members of the household should also be interviewed to capture generational and other differences among women who contribute to the family enterprise. Focus groups, participant observation, resource mapping, and other forms of qualitative data collection proved invaluable in interpreting quantitative results.[1] Various forms of institutional analysis also contributed to an understanding of women's roles in IPM.

References

Asturias, Linda, Sarah Hamilton, et al. 1999. Economic and socioeconomic impact assessment of non-traditional crop production strategies in small farm households in Guatemala. 1999. *Sixth Annual Report of the Integrated Pest Management Collaborative Research Support Program (IPM CRSP)*, pp. 149-255. Blacksburg, Va.: IPM CRSP.

Barrera, Victor, Jeff Alwang, María Crizón, Luis Escudero, Patricio Espinosa, Donald Cole, and George Norton. 2001. Intra-household resource dynamics and adoption of pest management practices. *Eighth Annual Report of the Integrated Pest Management Collaborative Research Support Program (IPM CRSP)*, pp. 406-412. Blacksburg, Va.: IPM CRSP.

Barrera, Victor, and George W. Norton. 1999. Farmer practices for managing the principal potato pests in the Province of Carchi, Ecuador: Results of a baseline survey and participatory appraisal. IPM CRSP Working Paper 99-1. Blacksburg, Va.: IPM CRSP.

Chrissman, Charles C., and Patricio Espinosa. 1993. Implementing a program of research on agricultural sustainability: An example of pesticide use in potato production in Ecuador. Draft paper presented at the Latin American Farming Systems Meetings, Quito, March 3-6.

Erbaugh, J. Mark, Joseph Donnermeyer, Magdalene Amujal, and Samuel Kyamanywa. 2003. The role of women in pest management decision making: A case study from Uganda. MS.

[1] There are several excellent guides to the collection of qualitative information during PAs that reveal the gendered dimensions of work, control of economic and natural resources, providership responsibilities, access to institutions, attitudes toward environmental issues, and economic and environmental goals. Among the best are those by Rachel Slocum et al., eds., 1995 and Barbara P. Thomas-Slayter et al., 1993 (see bibliography).

Hamilton, Sarah. 1998. *The Two-Headed Household: Gender and Rural Development in the Ecuadorean Andes*. Pittsburgh: University of Pittsburgh Press.

Hamilton, Sarah. 2000. The myth of the masculine market: Gender and agricultural commercialization in the Ecuadorean Andes. Pp. 65-87 in *Commercial Ventures and Women Farmers: Increasing Food Security in Developing Countries*, ed. Anita Spring. Boulder, Colo.: Lynne Reinner Publishers.

Hamilton, Sarah, Linda Asturias de Barrios, and Glenn Sullivan. 2002. Nontraditional agricultural export production on small farms in highland Guatemala: Long-term socioeconomic and environmental impacts. IPM CRSP Working Paper 02-1 (May 2002). Blacksburg, Va.: IPM CRSP.

Hamilton, Sarah, Linda Asturias de Barrios, and Brenda Tevalán. 2001. Gender and commercial agriculture in Ecuador and Guatemala. *Culture and Agriculture* 23 (3): 1-12.

Hamilton, Sarah, Gary Schlosser, Tina Schlosser, and Larry Grossman. 1999. Social and gender-related issues that affect IPM adoption in Jamaica. *Sixth Annual Report of the Integrated Pest Management Collaborative Research Support Program (IPM CRSP)*, pp. 331-335. Blacksburg, Va.: IPM CRSP.

Hamilton, Sarah, Glenn Sullivan, Linda Asturias, and Guillermo Sánchez. 2000. Economic and socioeconomic impact assessment of non-traditional crop production strategies on small farm households in Guatemala. 2000. *Seventh Annual Report of the Integrated Pest Management Collaborative Research Support Program (IPM CRSP)*, pp. 138-143. Blacksburg, Va.: IPM CRSP.

Harris, Colette, Danilo Vera, Miriam Cabanilla, José Cedeño, Victor Barrera, Jovanni Suquillo, María Crizón, Carmen Suárez, George Norton, Jeff Alwang, Luis Escudero, Cristina Silva, Byron Fonseca, Gerardo Heredia, I. Carranza, C. Belezaca, and R. Rivera. 2002. Intrahousehold resource dynamics and adoption of pest management practices. *Ninth Annual Report of the Integrated Pest Management Collaborative Research Support Program (IPM CRSP)*, pp. 324-329. Blacksburg, Va.: IPM CRSP.

Rocheleau, Dianne, Barbara Thomas-Slayter, and Esther Wangari, eds. 1996. *Feminist Political Ecology*. New York: Routledge.

Sachs, Carolyn E., ed. 1997. *Women Working in the Environment*. Washington, D.C.: Taylor & Francis.

Sachs, Carolyn E. 1996. *Gendered Fields: Rural Women, Agriculture, and Environment*. Boulder, Colo.: Westview Press.

Sissoko, Haoua Traoré, Anthony Yeboah, Keith M. Moore, Kadiatou Touré Gamby, Penda Sissoko Sow, Moussa Ndiaye, Alfousseini Ba, John Caldwell, Issa Sidibé, and Mah Koné Diallo. 2001. Gender analysis of IPM knowledge, attitudes and practices in horticulture. *Eighth Annual Report of the Integrated Pest Management Collaborative Research Support Program (IPM CRSP)*, pp. 184-187. Blacksburg, Va.: IPM CRSP.

Sissoko, Haoua Traoré, Colette Harris, Keith M. Moore, Mah Koné Diallo, Néné Coulibaly Traoré, and Issa Sidibé. 2002. Promotion of pest management practices by female farmers. *Ninth Annual Report of the Integrated Pest Management Collaborative Research Support Program (IPM CRSP)*, pp. 244-247. Blacksburg, Va.: IPM CRSP.

Slocum, Rachel, Lori Wichart, Dianne Rocheleau, and Barbara P. Thomas-Slayter, eds. 1995. *Power, Process, and Participation: Tools for Change*. London: Intermediate Technology Publications.

Tanzo, Irene, T. Paris, K.L. Heong, and Sarah Hamilton. 1998. Social impact assessment of IPM CRSP technologies in Central Luzon: Gender roles and intrahousehold decision processes in relation to rice-onion pest management in Nueva Ecija, Philippines. *Fifth Annual Report of the Integrated Pest Management Collaborative Research Support Program (IPM CRSP)*, pp. 275-277. Blacksburg, Va.: IPM CRSP.

Thomas-Slayter, Barbara P., Andrea Lee Esser, and M. Dale Shields. 1993. *Tools of Gender Analysis: A Guide to Field Methods for Bringing Gender into Sustainable Resource Management*. Worcester, Mass.: Clark University.

— V —
Conclusions

— 15 —
Lessons Learned

E. A. Heinrichs and S. K. De Datta

Introduction

Entering the 1990s, relatively few IPM programs in developing countries focused on fruits and vegetables. Most IPM programs emphasized cereal, especially rice, IPM for basic food security. Over the past 15 years, several small efforts and three major ones have been initiated or expanded that emphasize horticultural IPM in developing countries. The major efforts have been led by the IPM CRSP, the FAO Global IPM Facility, and the IARCs, especially AVRDC and CIP. Each of these efforts have taken participatory approaches; the IPM CRSP and the IARCs emphasized IPM research with linkages to national and NGO-based technology transfer programs, and the FAO program emphasized participatory IPM diffusion through FFS programs. The IPM CRSP has followed a consortium approach, with strong upstream and downstream linkages. Through experiences in Africa, Asia, Latin America, the Caribbean, and Eastern Europe, many lessons were learned on the CRSP about managing, program planning, technology development, and technology transfer. This chapter summarizes many of the key lessons learned.

The development and promotion of IPM depends on both technology development and on changing attitudes of farmers, scientists, extension workers, government officials, bankers, industry, NGOs, and consumers. IPM requires a common understanding among diverse stakeholder groups of the need for both productivity improvement and for attention to health and environmental concerns. Participatory approaches to IPM research and technology transfer are needed to build this understanding. Participation plays a major role in networking, private sector interaction, institution building, research-technology development, technology transfer, regionalization and globalization of IPM, change in government policy, and recognition of the importance of social and gender issues.

Participatory IPM (PIPM)

The development of PIPM has resulted in numerous lessons for effectively implementing IPM programs. PIPM means learning from, and with farmers, community members, policy makers, and others in order to investigate and evaluate constraints and opportunities. Lessons learned are used in the development and implementation of sustainable IPM programs.

A participatory approach is more demanding than a typical on-station approach. Successful development of PIPM requires an understanding of agricultural systems and involvement of all stakeholders in the chain from the producer to the consumer. Successful IPM programs require interactions among scientists, public and private extension, farmers, policy makers, and others. Lessons learned about the process of cost-effective participation in IPM are perhaps the most important contribution of the IPM CRSP to date. One measure of degree of success of this contribution can be found in the answer to the question: "Did the program lead to development of IPM technologies that changed behavior, met the production needs of small-scale farmers (was adopted), and reduced pesticide use?" The short answer is yes, and the long answer is partially, with significantly more progress needed in extending PIPM programs to other regions, countries, and agricultural systems.

The participatory approach

The participatory approach is based on the cooperation and participation of all stakeholders (see Chapters 3 to 7). Success of a participatory approach is determined in part by the extent to which it is able to link and enhance three key activities:

- **Problem identification**. Problem specification activities involve farmer surveys, participatory appraisals, crop-pest monitoring, collection of regional data, and planning workshops.
- **Research**. Research activities focus on biological, technical, and socioeconomic issues associated with a particular pest management problem, including long-term studies on the factors that influence pest populations. Implementation includes applied research projects, on-farm trials, and impact evaluations.
- **Communication, extension, and training activities**. Communication, extension, and training activities include conventional activities such as farm visits, field days, and demonstration plots, as well as more intensive efforts such as Farmer Field Schools where resources permit.

Initial participatory IPM activities include (1) stakeholder meetings, (2) participatory appraisals (PAs), and (3) baseline surveys (see Chapter 2 for details). The IPM CRSP experience has shown that successful participatory IPM (1) provides a solid scientific basis for the research program, (2) identifies pests, (3) provides possible solutions to pest problems, and (4) facilitates the spread of IPM strategies. The importance of the proper identification of pests is illustrated by the hot pepper-gall midge complex in Jamaica (Chapter 6), which caused significant export problems but was difficult to identify and thus delayed the development of management strategies.

Through the IPM CRSP experience in developing and transferring technologies in host countries, it was learned that the participatory research process requires researchers to consistently follow-up and follow-through with farmers. This need for frequent contact implies that on-farm research sites should be located close to the research stations so that scientists can handle the logistics in a reasonable manner. Once promising technologies are available, regional testing sites can be established on farms in the production centers in the country where contacts between researchers and farmers can be less frequent. Farmers know a lot about pests and pest management but not everything, and they look to scientists and extension workers for guidance in both pest-identification and innovative solutions. Farmers know more about weeds and insects than they do about diseases and nematodes. It is vitally important to include women in all aspects of the participatory process.

The participatory approach is an effective method of IPM technology development and transfer, but scaling up IPM research results to large numbers of farmers requires a combination of diffusion methods that simultaneously act to increase awareness (through methods such as mass media), take advantage of the private sector (especially for technologies imbedded in products in such as seeds, seedlings, bio-rationals, and pheromone traps), train trainers (through intensive methods such as FFSs), and achieve inexpensive spread of information (through methods such as demonstrations and field days). Additional modifications of the participatory approaches to IPM diffusion are needed to make them yet more cost-effective.

Networking

Strong networks are a basic element in a successful PIPM approach. Participatory IPM involves as many stakeholders as possible, and a mechanism that provides for that participation is networking. Table 15-1 lists the various groups, institutions, and organizations involved in technology

Table 15-1. Partnerships at regional sites (IPM CRSP, 2002-2003)

Site	U.S. universities	Host country government agencies	IARCs/ NGOs/ Coops/CRSPS
Southeast Asia–Philippines	Ohio State, VA Tech, Penn State	PhilRice, Inst. of Plant Breeding, National Crop Protection Center (NCPC) , Central Luzon State Univ., Leyte State Univ., Univ.of the Philippines, Los Baños (UPLB)	AVRDC, IRRI, NOGROCOMA
South Asia–Bangladesh	Ohio State, VA Tech, Penn State, UC-Davis, Purdue	Bangladesh Agricultural Research Inst., Bangladesh Rice Research Inst., Bongabandhu Sheikh Mujibur Rahman Agri. Univ., NCPC, UPLB	AVRDC, IRRI, CARE
East Africa–Uganda	Ohio State, VA Tech, Fort Valley State	Makerere Univ., KARI, NARO/NARI, SAARI, Pallisa DAO, Grain Crops Inst., Kumi DAO, Kaberamaido DAO, Iganga DAO, Nat. Ag. Advisory Services, Coffee Res. Inst.	CIMMYT, Quality Seed Co., ICIPE
West Africa–Mali	VA Tech, NC A&T, Montana State, UC-Davis, Purdue	Institut d'Economie Rurale (IER), OHVN, DGRC, LCV, Univ. de Mali	World Vision, Mali Primeurs Exporters, Flex Mali Exporters
South America–Ecuador	VA Tech, Ohio State, Florida A&M, Georgia	INIAP, FORTIPAPA, PROEXANT, Eco-Salud, Fundacion Maquipucuna, PUCE-IRD-Quito, ESPE-Quito, ESPOCH, MAG-Carchi	CIP, FAO, IFPRI, Soils CRSP, Vicosa Univ., Brazil
Central America–Guatemala/ Honduras	Purdue, Arizona, Georgia, Denver	ICADA, ICTA, FHIA, EAP-Zamorano, Univ. del Vallee, Guatemala, AGEXPRONT	ESTUDIO 1360, APHIS, FRUTESA, Transcafe

Table 15-1. Partnerships at regional sites, continued

Site	U.S. universities	Host country government agencies	IARCs/ NGOs/ Coops/CRSPS
Caribbean–Jamaica	Va Tech, Purdue, Ohio State, Penn State, UC-Davis, USDA-ARS	CARDI- Jamaica, St Kitts, St Vincent & the Grenadines, RADA, MINAG- Jamaica, Min. Agric & Fisheries-Trinidad, Food Storage & Prevention of Infestation Div. Univ. of West Indies- Mona	USDA/APHIS, CIPMNET, FAO
East Europe–Albania	VA Tech, Penn State, UC-Riverside, UC- Kearney Ag. Center	Plant Protection Institute, Agric. Univ. of Tirana, Fruit Tree Res. Institute, MoA Directory of Science and Extension Service	Albanian Organic Agriculture Association, Alimentary Oil Association

development and transfer at each of the IPM CRSP global sites in 2002-2003. This list does not include all of the institutions such as banks, consumer groups, private industry, etc. that are also part of the overall network and have an input into the direction of IPM programs.

The network approach provides a pool of expertise to meet unique problems at each site related to technology development, technology transfer, gender issues, policy instruments, and export and quarantine problems. Having U.S. universities, host country partners, IARCs, NGOs, and others working at each site provides the range of necessary disciplinary expertise (Figure 15-1). The makeup of the multi-institutional teams differs from site to site depending on the constraints. In Jamaica, for example, where hot pepper exports are a problem, the teams include agencies involved in the hot-pepper export industry. The networks have been a major reason for the success of the IPM programs at each regional site. In Ecuador, for example, linkages among INIAP, FORTIPAPA, PROEXANT, Eco-Salud, Fundacion Maquipucuna, PUCE-IRD-Quito, ESPE-Quito, CIP, FAO, IFPRI, Soils CRSP, Vicosa Univ. (Brazil), ESPOCH, MAG-Carchi, PROMSA (World Bank ag technology and training project), and other agencies strengthened the pool of expertise in support of project objectives. In Albania, the Participatory IPM approach helped move the key agencies from a lack of cooperation to excellent cooperation, resulting in successful development and transfer of Olive IPM technology. In all of these examples each partner

Figure 15-1. A subset of IPM CRSP U.S. and host country collaborators in Bangladesh.

played a key part in the IPM programs, and success would not have been possible had only one agency been involved.

Private Sector Involvement

Private sector involvement is critical for certain aspects of a PIPM program. For those commodities aimed at the export market, the involvement of export-marketing cooperatives or firms can help guide the IPM program in the direction of meeting the requirements of an unforgiving market. For several component technologies — such as insect- or disease-resistant seeds, pheromone traps, and grafted seedlings — the private sector is essential to ensuring that key inputs do not become a constraint to technology adoption. The private sector can also be a hindrance to IPM to the extent that output-marketing firms may concentrate primarily on large farms. In addition, private pesticide salespersons may attempt to discourage use of non-chemical means of pest management. However, where there is mutual interest between the public and private sectors, cooperation should be encouraged.

Small private cooperatives have been important to the success of IPM CRSP programs. In the Philippines, the national onion cooperative, NOGROCOMA, has been instrumental in technology transfer. They represent most of the onion farmers in the country, and have worked closely with the CRSP in testing component technologies. In Guatemala, fruit and vegetable cooperatives such as Cuatropiños have helped farmers to achieve the level of quality control required in export markets by transferring IPM knowledge to growers and by developing pre-inspection protocols for export crops.

Non-Governmental Organizations (NGOs)

NGOs make use of a combination of public and private funds to reach small farmers and to address management practices ignored by the private sector. Several NGOs have had a major impact on the success of the IPM CRSP program in certain sites. For example, CARE Bangladesh has collaborated with the IPM CRSP; the primary objective is promoting appropriate IPM technologies for vegetables for marginal farmers, through its Rural Livelihood Program. This collaboration has been effective in improving production levels and raising incomes for marginal farmers. With CARE's long experience in working with the rural poor in Bangladesh, they have been able to transfer vegetable IPM technologies developed by the IPM CRSP to a much wider audience than would have otherwise been possible. Through their long-term involvement in a country, NGOs and the private sector can continue IPM diffusion programs beyond the life of the IPM CRSP; thus, with their involvement, IPM activities become more sustainable .

Institutional Capacity Building

The development of strong institutions that can continue the design and transfer of IPM technologies after donor support from projects such as the IPM CRSP leave is critical to globalizing IPM. The training of scientists (and others) is a key component in building IPM capacity in local institutions. Capacity building also involves giving an identity and visibility to IPM programs within each country so that they are appreciated and supported by the countries themselves. It involves establishing methods for scientific planning, participatory conduct, review, and evaluation of IPM research and technology transfer so that these activities are recognized as being of high quality and relevant. Dr. Santiago Obien, former executive director of Phil Rice in the Philippines, said that a major contribution of the IPM CRSP in his institution was the effect of institutionalizing the process of IPM research planning, conduct, and review.

Institution building involves a mix of training, both long- and short-term. Long-term training means graduate degree-training. During the first 11 years of the IPM CRSP, 93 students received graduate degrees through the IPM CRSP. The vast majority of this training was for host-country scientists (75%) and the remainder for American students. Training of host-country scientists included training in their country, another country in the region, or the United States. Some students participated in a "sandwich" program through which a portion of the program (courses and/or research) was conducted in the host country and the other portion in the United States or in an institution outside the country such as at AVRDC. Sandwich programs are less expensive and take scientists out of their systems for a shorter period of time than, say, a Ph.D. in the United States. Distance Education will become more important in the future.

An example of institution building is the IPM CRSP effort in Mali to build up the capability of the Environmental Toxicology Laboratory (ETL), which had been built and staffed under an earlier project but was not functioning at a level needed to support either the domestic or the export program. Before the CRSP involvement, no scientists at the ETL were specifically trained in the environmental monitoring of water and soil, an area of expertise important in a country with minimal enforcement of pesticide markets and large-scale storage of obsolete pesticide stocks. Consequently, the USAID/Mali Mission requested the IPM CRSP to transform the laboratory infrastructure into a functioning unit. The CRSP established an overall laboratory-development and staff-training program to develop capacity to monitor pesticide residues. Short-term training provided the initial impetus for this development. But for sustainability, long-term training was needed. Thus, an ETL technician was enrolled to pursue a graduate degree at Virginia Tech. In May 2004, she completed an M.S. degree in Environmental Toxicology. Having advanced her knowledge of pesticide residue analysis through graduate field and laboratory studies, she returned to the ETL of the Laboratoire Central Vétérinaire in Bamako, Mali, with a wide range of analytical skills. With the addition of this highly trained scientist, the range of residue analyses that the ETL could conduct increased significantly. The training led to sustainable functioning of the pesticide residue laboratory beyond the IPM CRSP project period. The scientific capacity will have far-reaching impacts on development of pesticide quality assurance in the region, promoting human and animal food, export crops, a healthier environment, and sustained economic growth.

The IPM CRSP has met the need for short-term training of scientists through training periods of 2 weeks to 6 months at U.S. universities, host-country universities, on-site in the host countries, and at the IARCs. Short-term training is used to increase the capacity to conduct specific research tasks, thus making scientists proficient in a given technique needed for their research program.

The IPM CRSP research in Central America has been instrumental in establishing and institutionalizing protocols supporting the exportation of snowpeas from Guatemala. The snowpea program was the first fully integrated program to be approved by the Government of Guatemala and certified by APHIS. This activity involved a large array of stakeholders. At this time, more than 40% of all snowpea production in Guatemala is produced and marketed under IPM CRSP-developed protocols. Thus the Guatemalan snowpea producer is assured greater access to U.S. markets and U.S. consumers are assured of safer food supplies. The IPM CRSP experience in Guatemala indicates that when current scientifically tested and proven production technologies are properly integrated and precisely managed, the production goals of immediate economic gain and long-term sustainability are mutually reinforcing.

Institutionalization of an IPM program in the fabric of a country takes considerable time; hence the need for a long-term horizon. Although some research projects can be completed in 3-4 years, a longer period is needed to institutionalize the IPM program and develop a sustained technology-development and -transfer process.

IPM Technology Development

IPM technology development must stress a close linkage between farmers and the research program. A systems approach can be used to integrate information of various types (technical, economic, climatic, biological, etc.) and lead to an understanding of pest-population dynamics, markets, and policy constraints. Developing IPM packages has involved the employment of multidisciplinary and multi-institutional teams, including most of the critical stakeholders. Certain crops require significant research prior to technology transfer. For example, extensive research has already been conducted on rice, and this information is ready to be transferred to farmers, whereas with vegetables, much of the needed information to develop IPM tactics is lacking. Thus, the IPM CRSP emphasized vegetable research prior to the transfer programs. Much of the research has been participatory and has been conducted in farmers' fields, but it also involves judicious use of variety screening on station as well as greenhouse experiments, where prudent, to speed up the research process and shorten the time from research to technology transfer to the farmer. This approach has proven effective at all IPM CRSP sites globally.

Technology Transfer

The ease of transferring technology depends on the environmental sensitivity of the technologies, and on environmental, cultural, and other sources of diversity within countries. To speed diffusion of IPM, a multifaceted approach is needed in which all agencies are utilized: (1) Traditional public extension agencies, (2) private for-profit firms, and (3) private non-profit entities. "One size fits all" does not work in IPM diffusion. Instead, a multifaceted approach is needed because of:

- Differing complexities of IPM technologies
- Differences in local public extension capabilities
- Differences in resources
- Differences in education
- Differences in socio-economic circumstances

Farmers' Field Schools (FFSs)

FFS programs have been used extensively to facilitate the spread of IPM technologies to farmers, especially through programs promoted by FAO but also on the IPM CRSP (See Chapter 9). Through this approach many valuable lessons have been learned:

1. FFSs need to be adapted to meet the needs and interests of each specific region/community.
2. FFSs must include a successful training-of-trainers (ToT) program to be cost-effective. Trainers must have appropriate IPM and methodological training and skill in managing participatory, discovery-based learning as well as a technical knowledge of agro-ecology.
3. Many solutions to agricultural development are not just *technological* in nature, but *conceptual* and social. FFSs can respond to both.
4. NGOs are valuable partners.
5. Special care is needed to avoid turning the learning field into a competition between farmers and facilitators.

6. In many places, the majority of the FFS participants are women. Women are often active participants in pest-management decisions and their involvement in IPM training is critical.
7. A current challenge is to establish collaborative structures and finance- and technical-support mechanisms to sustain the FFS movement. Average costs of $40-50 to train one farmer are too high for most countries. FFSs must be modified or combined with more cost-effective methods. FFSs continue to reach only a small portion of farmers and communities. The solution includes:
   - Improving the flow of information and technology from the FFS participants to non-participants.
   - Further developing self-financing opportunities for FFS programs.
   - Complementing FFS with mass media and other approaches to diffusing IPM knowledge.
   - IT fact sheets in different languages, radio programs, comic strips, etc. to enhance awareness and to promote diffusion of IPM technologies.

Regional spread of IPM technology

To increase the broader regional impact of IPM research, donor-supported IPM programs such as the IPM CRSP must select crops, pests, and technologies of regional significance. For example, eggplant is the most important vegetable in many countries in Asia, and fruit and shoot borer and bacterial wilt are common pest problems. Grafting of eggplant varieties susceptible to bacterial wilt onto resistant rootstocks, a practice perfected in Bangladesh with the help of AVRDC and the IPM CRSP, has potential to spread within Bangladesh and across the region. Transferring this profitable eggplant-grafting technology to millions of Bangladeshi vegetable farmers is difficult, but not impossible. The IPM CRSP-BARI in June 2003 teamed up with CARE-Bangladesh for technical exchange and support to elevate the livelihoods of vegetable farmers and alleviate poverty in Bangladesh. The eggplant-grafting technology is being transferred to thousands of vegetable farmers through this collaborative effort. Furthermore, private sector nurserymen are being trained to do the grafting and sell the seedlings to farmers. In addition, women in individual villages are being trained to do the grafting and they too are selling seedlings.

Regionalization to other countries in Asia beyond Bangladesh for this technology has involved sending scientists from other countries to

Bangladesh to learn how to do the grafting. For example, the Philippines sent two scientists to learn the grafting technique from Bangladeshi scientists. Grafted eggplants are now also being produced in the Philippines and should have a significant impact on the economics of eggplant production there as well.

Non-economic constraints to trade represent potentially challenging limitations to future regional development in the NTAE (Non Traditional Ag Exports) sector in Central America. The integrated crop-management strategy in snowpea developed in Guatemala serves now as a model for achieving sustainable (NTAE) production in other countries in the region (Sullivan et al., 2000).

Regional workshops and symposia are also important mechanisms for spreading IPM information and technologies. Scientific papers are critical, as are databases, websites, and newsletters that record and publicize IPM successes that may apply in other sites.

Government Policies

Government policies can encourage or discourage the development and adoption of IPM technologies. Thus, policy analysis is often an integral part of a successful IPM program. If policies create barriers to IPM adoption, such that there is little economic incentive to adopt, there may be little return to IPM technology development and transfer. The establishment of policies supporting the economic incentives of IPM practices, vs. only applying pesticides, is critical to the success of IPM programs.

Regulations, taxes, subsidies, cooperative organizations to control pests in a region, laws and policies that influence land tenure, cooperative marketing schemes, credit policies, and other policies are difficult to change. Yet the incentives created by these "rules of the game," and the market barriers created if they are ignored, can spell the difference between successful and unsuccessful IPM efforts.

As indicated in Chapter 10, the policy issue of defining the appropriate mechanism for diffusing results of IPM research is also of vital importance but is in its infancy. Several diffusion approaches have been utilized, but evidence of the appropriateness and economics of the different approaches in various settings is minimal. A key to successful IPM programs is to design and support workable, cost-effective approaches to IPM. Thus, an understanding of the policy issues affecting IPM is basic to successful IPM programs. There must be significant policy dialogue among policy makers and economists who are engaged in policy research. The most difficult task is often not the analysis of the policies and institutions but the policy dialogue that must occur to convince governments and others of the need for change.

Gender Issues

The farmer is often thought of as male but, in practice, a high percentage of women are farmers as well. In sub-Saharan Africa, women perform most of the agricultural labor (Erbaugh et al., 2003). Gender differences that can influence adoption and impacts of IPM include (1) ownership and control of resources, (2) family provisioning responsibilities, (2) division of household labor, (4) specialized knowledge of crops, pests, and traditional pest-management practices, (5) attitudes toward health consequences of pesticide use, and (6) social roles (Bonabana-Wabbi, 2003). Gender is culture specific, which means it is modulated by ethnicity, social class, economic and educational levels, community, age, and so forth.

IPM CRSP research in Mali found that when Farmer Field Schools were limited to men, women were more likely than men were to believe that it is necessary to use chemical pesticides; they applied pesticides to their plots in the same fields where their husbands were using alternative IPM techniques. Apparently, knowledge learned through FFS does not spread well within Malian households (another illustration of why FFS is an expensive tool for technology diffusion).

It is not only in Mali that women play a significant role in pest management. Studies conducted at IPM CRSP sites in Uganda, Albania, Guatemala, Ecuador, the Philippines, Bangladesh, and elsewhere found that women have an important role in pest management. In the IPM CRSP, a special effort has been made to not only include women farmers wherever possible in all technological-transfer activities, but also to use women scientists and women extension agents as much as possible.

Conclusions

The development and transfer of IPM technologies is a complex and dynamic process. New pest constraints and new approaches to solving those constraints are constantly being developed. A key to successful IPM programs is the participation of all stakeholders. The final evidence of the success of IPM programs is the extent of farmer adoption, or adaptation of the IPM technology. The IPM CRSP sites are witnessing growing adoption of its IPM strategies despite the fact that the program emphasis has been on participatory research as opposed to technology transfer *per se*. Participation in the program of technology-transfer organizations such as NGOs and national extension services is one reason why. One lesson that has been learned is that one size does not fit all when it comes to IPM technology development and transfer, due to the site specificity of economic, social, institutional, and agro-ecological factors. Therefore, a participatory approach that follows a set of basic principles, as described in Chapter 2, is perhaps the best way to ensure globalization of IPM.

References

Bonabana-Wabbi, J. 2003. Assessing factors affecting adoption of agricultural technologies; the case of Integrated Pest Management (IPM) in Kumi district, Eastern Uganda. M.S. thesis, Virginia Tech. Blacksburg, Va.

Erbaugh, J., M. Amujal, S. Kyamanywa, and E. Adipala. 2003. The role of women in pest management decision making in Eastern Uganda. *Journal of International Agricultural and Extension Education* 10-3.

Sullivan, G.H., S.C. Weller, C.R. Edwards, P.P. Lamport, and G.E. Sánchez. 2000. Integrated Crop Management strategies in snowpea: A model for achieving sustainable NTAE production in Central America. *Sustainable Development International*, Autumn 2000, pp. 107-110.

Index

A

B

C

D

E

F

G

H

I

M

N

O

P

Q

R

U

V

W

X

Y

Z